Isles of Refuge

Laysan Albatross sails toward Nihoa.

Isles of Refuge

Wildlife and History of the Northwestern Hawaiian Islands

Mark J. Rauzon

A Latitude 20 Book

University of Hawai'i Press

Honolulu

Published with the support of the Maurice J. Sullivan
& Family Fund in the University of Hawai'i Foundation

06   05   04   03   02   01   1   2   3   4   5

Library of Congress Cataloging-in-Publication Data
Rauzon, Mark J.
Isles of refuge : wildlife and history of the northwestern Hawaiian Islands /
Mark J. Rauzon.
p.   cm.
"A Latitude 20 book."
Includes bibliographical references (p.).
ISBN 0–8248–2209–9 (cloth : alk. paper) —
ISBN 0–8248–2330–3 (pbk. : alk. paper)
1. Zoology—Hawaii.   2. Hawaii—History.   I. Title.
QL345.H3 R38 2001
591.9969'9—dc21   00–029874

All photographs and illustrations are by author unless otherwise noted.

University of Hawai'i Press books are printed on acid-free paper
and meet the guidelines for permanence and durability of the
Council on Library Resources.

Designed by Nina Lisowski

Printed by Friesens

Contents

Acknowledgments

This book is dedicated to Sheila Conant, Dave Woodside, and Karl Kenyon, who taught me firsthand about the Leewards. I am especially grateful to the U.S. Fish and Wildlife Service (USFWS) for access, companionship, and support in the Northwestern Hawaiian Islands. The biotechnicians of the USFWS, as well as those of the National Marine Fisheries Service (NMFS), whose data collection is the basis of most management decisions that affect wildlife, are an inspiration. I also thank the crew of the National Oceanic and Atmospheric Administration ship *Townsend Cromwell* for transportation and logistical support, without which this book could not have been written. Mahalo to Hui Mālama and the Polynesian Voyaging Society for an opportunity to sail with them. I am particularly grateful for the companionship of Kimo Lyman and Nainoa Thompson of the *Hōkūle'a*.

I am indebted to Alan C. Ziegler for his thorough technical review, to Eileen D'Araujo for her expert copyediting, and to Suzanne Rauzon and Sheila Conant for their helpful editorial comments. The spirit of Lani Stemmerman inspired me to complete this work. Bobby Camara was her medium. I am grateful to Beth Flint, Eric Knudtson, Karla Kral, and Susan McKinley for support and encouragement. Craig Harrison, Skip Naftel, and Sue Scatolini provided timely insights. I thank Jason Baker, Chris Boggs, Brenda Becker, and John Henderson of NMFS; Ken McDermott, Craig Rowland, and Nanette Seto of USFWS; and Diane Drigot, U.S. Marine Corps, for help and information. Roger Clapp, Storrs Olson, and other Smithsonian Institution scientists who coauthored the Atoll Research Bulletins presenting historical and biological data from the Leewards are the shoulders I stand on. Thanks to John Gilardi and Maura Naughton for illustrations.

Helene Schlemmer Brown allowed me gracious access to her family history. The late Ed Bryan of the Pacific Scientific Information Center, which he managed for the Bishop Museum Archives, and his books about the Leewards were a guiding light. The late Keneti, Kenneth P. Emory, also of the Bishop Museum, provided insights to my Necker questions. Librarians at the Bishop Museum Archives, the Hawai'i State Archives, and the U.S. National Archives, Pacific Division, were extremely helpful. The Denver Museum of Natural History also contributed extinct Laysan land bird photos.

The poem at the end, "XXIV-The Island," from *The Separate Rose* by Pablo Neruda, translated by William O'Daly (1985), is reprinted with permission of the Copper Canyon Press, P.O. Box 271, Port Townsend, Washington 98368. The Hawaiian sayings throughout, unless otherwise noted, are from *'Ōlelo No'eau* by Mary Kawena Pukui (1983), Bishop Museum Press.

Mahalo and aloha.

Pronunciation of Hawaiian Words

Vowels

A as in **o**tter.
E as in h**ay**.
I as in **ea**t.
O as in g**oa**t.
U as in h**oo**t.

Consonants

H, L, M, N: approximately as in English.
K, P: approximately as in English, but with less air released.
W may be pronounced as w or v: usually v after i and e, and usually w after u and o.
' (glottal stop) is like the sound between the ohs in English oh-oh.

A macron indicates a long, stressed vowel.

Accent is on the penultimate vowel if all vowels are short, on diphthongs (ai, au, ua, etc.), or on stressed vowels (with macron).

Introduction: The Northwestern Hawaiian Islands

Ua mau ke ea o ka 'āina i ka pono.

"The life of the land is perpetuated in righteousness." Hawai'i State Motto

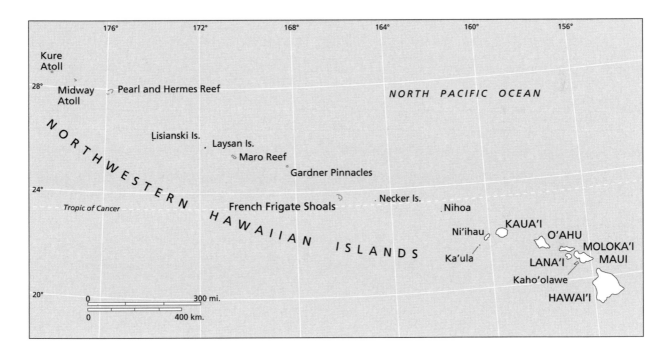

Map of the Hawaiian Archipelago and Emperor Seamounts.

Hawai'i's state motto sounds eloquent and lyrical, but in reality the native life of the land is threatened with extinction, crowded out by organisms from every part of the world. The coastline is overfished, the land is dense with development, and the burgeoning human population challenges the environment at every turn. How-ever, in the uninhabited Northwestern Hawaiian Islands, righteousness reigns. The entire range of marine life originally found in the main islands is still largely intact. Beach creatures such as monk seals, sea turtles, and seabirds continue to thrive in a marine environment unlike any elsewhere in Hawai'i. As a wildlife biologist with the U.S. Fish and Wildlife Service, I visited all but one of these remote islands to study the exuberance of life that is Hawai'i's heritage.

The Hawaiian Archipelago appears to be divided into two halves: the eight main or high islands clustered in the southeast and the ten far-flung isles and atolls of the northwest. Arcing 1,100 miles across the middle of the North Pacific Ocean, the Northwestern Hawaiian Islands are each about 150 miles apart. The ten are collectively called the Leeward Islands, even though they are not in the lee of the high islands and have been exposed to the wind and rain for millions of years longer than the main islands. The elements have reduced these once-massive volcanoes to only about 3,328 acres of dry land or about one-tenth of 1 percent of the land area in the entire Hawaiian Archipelago.[1]

Satellite image of the main Hawaiian Islands. (NASA photo)

The Northwestern Hawaiian Islands are so small that they rarely appear on tourist maps of Hawai'i and are virtually unknown to many visitors and Hawaiian residents. Only Midway Atoll, site of a definitive World War II battle, is recognized by most people. Approximately 250 miles from Honolulu, the southernmost Leeward island is Nihoa, followed to the northwest by Necker Island, French Frigate Shoals, Gardner Pinnacles, Maro Reef, Laysan Island, Lisianski Island, Pearl and Hermes Reef, Midway Atoll, and finally Kure Atoll.

All Hawaiian Islands had a fiery birth. Molten lava seethed out from a submarine vent in the earth's crust at a site called a "hot spot." Unceasing eruptions for thousands of years built mountains that rose as high as 15,000 feet above sea level. Including their underwater bases, the volcanoes of the island of Hawai'i, the Big Island, are considered to be the highest mountains on Earth, reaching over 35,000 feet from their base on the sea floor. The "hot spot" that spawned all the islands in the Hawaiian Archipelago is now located 20,000 feet below sea level, slightly southeast of the Big Island. The next Hawaiian Island under construction, called Lō'ihi, is growing 3,000 feet beneath the sea.

It is only in the last 30 years that we are beginning to fully understand the geology of Hawai'i. William Alanson Bryan, professor at the College of Hawaii from 1909 to 1919, was a believer in the theory that at one time there was an immense continent in the Pacific that sank below the surface of the water eons ago, leaving the highest mountaintops as islands.[2] Modern science has proved the esteemed scientist wrong. Volcanism, continental drift, erosion, and coral growth, coupled with changes in sea level and periodic subsidence and reemergence of islands during past periods of global temperature fluctuation, have created a linear set of islands and seamounts in various stages of development. A chain of submarine volcanoes, known as the Hawaiian Ridge, includes all the islands of the Hawaiian Archipelago as well as numerous seamounts that do not reach the ocean surface. All the volcanoes rest on the portion of the earth's crust called the Pacific Plate. This section of the earth's crust is moving in a northwesterly direction, at the rate of about three and a half inches a year.[3] Each island or seamount along the entire Hawaiian Archipelago was formed when the Pacific Plate passed over the same stationary fountain of molten rock, the "hot spot."

Lava entering the ocean at Kalapana, Hawai'i Volcanoes National Park.

After an island is pulled away from the magma source, the volcanic eruptions cease and erosion begins to reclaim the land. Trade winds, blowing consistently out of the northeast, and associated rainfall have sculpted great fluted valleys into the flanks of most islands. In addition, as an island "melts" in the rain, the great weight of the mountain of lava causes the crustal plate to sag. The island subsides in the sea and this speeds erosion above sea level. But in the warm waters that bathe Hawai'i, an island again begins to grow from below sea level. Calcium carbonate accretions by coral animals build up like a crown on sinking volcanoes, creating fringing reefs that break the relentless waves. Coral reefs help slow erosion, and coralline algae growth helps cement the erosional debris together, keeping the island from slipping beneath the sea. The Hawaiian Archipelago would be five islands and approximately 750 miles shorter were it not for this protection.[4]

It was Charles Darwin who first elucidated the theory of atoll development in 1842. Much later, geologists confirmed his theory when they drilled through the coral cap and hit the underlying basalt.[5] Core samples at Kure Atoll, the oldest Hawaiian island, indicate that it was over the "hot spot" about 60 million years ago. The highest-latitude coral reef in the world, Kure Atoll is across the so-called Darwin's Point: where coral growth is surpassed by island subsidence. Darwin's Point has remained within two degrees of 29° N for 20 million years.[6] When coral growth at Kure cannot keep pace with the rate of subsidence, the low-lying atoll will eventually sink beneath the surface. This fate awaits all the Hawaiian Islands in geologic time.

Witness the baker's dozen of now-drowned islands that extend beyond Kure. Called the Emperor Seamounts and named after Japanese rulers, these defunct Hawaiian Islands turn northward and stud the abyssal ocean plain up to the Aleutian Islands of Alaska. The seamounts, also called guyots, begin at the Hancock Seamount near Kure Atoll and end with the oldest surviving seamount, Meiji Guyot—some 2,000 miles away. The entire Hawaiian Ridge–Emperor Seamount Archipelago is the result of at least 100 million years of virtually nonstop volcanic activity. Like a great geological conveyor belt, the Pacific Plate is being subducted, or dragged under the Alaskan crustal plate. Hawaiian Islands are recycled in a cauldron of magma underneath this portion of the "Ring of Fire" that borders the Pacific Ocean basin.

The most isolated fleet of islands in the world, the Hawaiian Archipelago is about 2,500 miles from the North American continent and 3,800 miles from Asia. Vast oceanic distances strictly limit the type and number of species that can survive the journey to its remote shores. Only those life forms able to swim, sail, fly, or float for thousands of miles ever reach these subtropical islands. Common continental species such as ants, frogs, snakes, land mammals (except bats), bamboo, coconuts, and pine trees never established themselves naturally. The few species that did survive the epic journey to Hawai'i underwent selective adaptation to take advantage of a multiplicity of life zones. A combination of rich volcanic soils, abundant water, and the hot sun of the tropic of Cancer spawned a splendid paradise of evolutionary creativity. From coral atolls to snowy mountaintops, from rain forests to lava deserts, colonizing animals and plants had a vast range of habitats to exploit with little predation or competition. Over millions of years, organisms evolved in unimaginable ways: carnivorous caterpillars, happy-faced spiders, giant geese, and flightless ibis, to name but a few.

Tree ferns growing in mist from volcanic steam vent, Hawai'i Volcanoes National Park.

About 1,000 years ago, this genetic experiment changed drastically when the winds carried prospecting Polynesians to Hawai'i. The voyagers brought pigs, dogs, rats, chickens, and their respective appetites to an archipelago where thornless plants and flightless birds flourished. The colonizers cut, burned, and cultivated the lowland forests, hunted the genetically tame wildlife, and released their domestic animals to forage on their own. Native species rapidly disappeared as human influence expanded. As a result of Polynesian land use, at least fifty species of endemic birds had been extirpated when Captain James Cook arrived in

Laysan Albatrosses engaged in mutual preening, Tern Island, French Frigate Shoals.

and sea turtles inhabiting the Leewards. President Theodore Roosevelt, in the waning days of his presidency, acknowledged the critical need to protect the birds from the ravages of humans. He issued Executive Order 1019 on 3 February 1909, creating the Hawaiian Islands Bird Reservation (one of fifty-one in the nation at that time). On 25 July 1940, President Franklin Delano Roosevelt changed the name of the bird reservation to the Hawaiian Islands National Wildlife Refuge. Today, under this legal provision, most of the animal and plant populations of the northwestern islands have recovered to preexploitation levels. Nine of these islands are included in the National Wildlife Refuge system, managed by the U.S. Fish and Wildlife Service (USFWS). Kure Atoll, the north-

1778.[7] During the next 200 years, the onslaught of destruction and biological invasions accelerated to biblical proportions. Another thirty-one species of birds have vanished and native ecosystems now cling to a tenuous existence. Hawai'i leads the world in the number of endangered species, whose last retreat is mostly on the island summits. Species that were adapted to lowland ecosystems have suffered the most in the main islands, but those of the uninhabited northwestern isles can still find refuge.

But it was not always the case. Even these tiny islands were not spared centuries of exploitation when wholesale slaughter was the rule. In the late nineteenth and early twentieth centuries, sealers, feather hunters, fishermen, and guano miners killed most of the seals, seabirds,

Hawaiian Monk Seal males sparring at Tern Island, French Frigate Shoals.

ernmost island in the chain, is a state wildlife sanctuary.

All the land in the Northwestern Hawaiian Islands falls under the civil jurisdiction of the State of Hawai'i and the City and County of Honolulu, except Midway Atoll, which is owned by the federal government and is thus not legally part of the state of Hawai'i. The U.S. Coast Guard patrols the 200-mile limit of the exclusive economic zone surrounding the northwestern islands. International shipping and fishing vessels as well as the recreational sailing public are forbidden to pass though this area. This "closed-door" policy is a USFWS regulation designed to quarantine the vulnerable island ecosystems from the inadvertent introduction of pests and weeds.

References

1. Armstrong (1983).
2. W. A. Bryan (1931).
3. Juvik and Juvik (1998).
4. Grigg and Dollar (1980).
5. Lobban and Schefter (1997).
6. Grigg and Dollar (1980).
7. Olson and James (1991).

1

Fathoming the Past

Kū pākū ka pali o Nihoa i ka makani.

"The cliffs of Nihoa stand in resistance to the wind."
Said of a person who stands bravely in the face of misfortune.

Infrared photo of Nihoa. (USFWS photo)

The first of the Northwestern Hawaiian Islands is Nihoa. Shaped by the same forces that carved the precipitous northwest cliffs of Kaua'i's famous Nāpali coast, 135 miles away, Nihoa's 900-foot cliffs jut straight up like the rugged ramparts of a fallen fortress. Almost whittled away by the relentless winds and rains, Nihoa stands on the southwest portion of what was the island's original volcanic cone. Approximately 7.2 million years of erosion have left only 156 rugged acres of this once-massive island sitting on an underwater base about 17 miles in diameter and 20 fathoms under the sea.[1] Little coral growth occurs at Nihoa because of the island's steep submarine slopes. Without reefs to break the waves, water continues to etch Nihoa. Within sea-sculpted lava cliffs, the figures and faces of *kupua*—fantastic mythical creatures of ancient Polynesia—seem to appear.

Nihoa is the geologic and cultural fulcrum for the two halves of the Hawaiian Archipelago. The tallest of the leeward islands, Nihoa was once inhabited by ancient Hawaiians, either purposeful colonizers or shipwrecked survivors, who negotiated the treacherous wave surge, the slippery lava benches, and the rugged cliffs to find refuge. They called the island Nihoa, "toothed" in Hawaiian.[2] The people survived by building houses into cliff faces and terracing the rocky soil to grow sweet potatoes. They had abundant fish and birds to complement their starchy diet, shelter from the elements, and they could appease their gods at the numerous religious sites that dot the island. At least several generations survived and left compelling evidence of their culture before abandoning the island. Radiocarbon dating of charcoal discovered on Nihoa provisionally suggests that the island was colonized around

A.D. 1000 and could have been occupied over a 700-year period.[3]

Today, Nihoa is vaguely known to many people simply as an island west of Kaua'i and is often confused with Ni'ihau, which lies within the rain shadow of Kaua'i, 17 miles to the southwest. Ni'ihau is called the Forbidden Isle because it is privately owned and off limits to visitors. It supports a population of about 230 Native Hawaiians who eschew modern life and preserve a remnant of Hawaiian culture living on the land or *'āina*.

Nihoa might be called the Forgotten Isle, because it was abandoned and virtually unknown when it was found by Westerners. The credit for rediscovery has historically gone to Captain William Douglas of the trading vessel *Iphigenia*, who sighted Nihoa at three o'clock on the morning of 19 March 1789.[4] But Douglas was a year late, because Captain Colnett, the commander of the *Prince of Wales*, discovered it on 21 March 1788.[5] A source no less than the great English explorer Captain George Vancouver wrote in his journal, *A Voyage of Discovery Round the World:*

It was not much out of our way to ascertain the situation of a small island, discovered in 1788, by the commander of the *Prince of Wales* and by him called Bird island, in consequence of its being the resort of vast flocks of the feathered tribe. . . . When this rock was first discovered in 1788, there were on board the *Prince of Wales* some natives of *Attowai* [Kaua'i] who expressed great surprise that there should be land so near to their islands . . . of which not only themselves, but all their countrymen were totally ignorant. This intelligence was communicated on their return in the autumn of

that year; and it excited in the active mind of Taio [King Kaio of Kaua'i], a strong desire to pay it a visit, to establish a colony there, and to annex it to his dominions; but on his being made thoroughly acquainted with its extent and sterility... his project was abandoned. Those people however recognize it under the apellation of *Modoo Mannoo,* this is, Bird island; and from its great distance from all other land, and its proximity to their islands, it seems to claim some distant pretensions to be ranked in the group of the Sandwich islands. . . .[6]

In March 1789, Captain Douglas and the *Iphigenia* left Kaua'i in search of Nihoa. The ambitious owner of the vessel, John Meares, directed Captain Douglas to claim Nihoa. Meares was the first to record its description: "This island, or rock, bears the form of a saddle, high at each end, and low in the middle. To the south, it is covered with verdure; but on the north, west and east sides it is a barren rock, perpendicularly steep, and does not appear to be accessible but to the feathery race,

Nihoa at sea.

with which it abounds. It was therefore named Bird Island."[7]

Meares took credit for Colnett's discovery of Nihoa as a way to boost the recognition of his name as an aspiring geographer and explorer. Captain Colnett was never able to contest Meares' usurpation because Colnett's ship was soon captured by the Spanish off the west coast of America. He went insane in prison and returned to England a broken man. Historian George Evering wrote, "Give him credit in the history books, but reserve the final accolades for those intrepid Hawaiian sailors who first raised this jagged isle from the sea. And that hundreds of years before the English dared leave sight of land."[8]

In spite of Vancouver's assertion that no one on Kaua'i knew of Nihoa, a reference to the island existed in ancient chants. While living on Kaua'i in 1822, Queen Ka'ahumanu, the widow of Kamehameha the Great (the Hawaiian chief who unified all the islands of Hawai'i), heard old *mele* or chants about the islands. Perhaps she heard the one by Kawela Mahunaalii:

Nihoa's perpendicular cliffs drop 600 feet below Pinnacle Peak.

Ea mai ana ke ao ua o Kona,	The rain cloud of the south comes,
Ea mai ana ma Niho-a,	Comes from Nihoa,
Ma ka mole mai o Lehua,	From beyond Lehua,
Ua iho a pulu ke kahakai.	Rain has flooded the beach.[9]

Perhaps Ka'ahumanu heard the most romantic-mystical aspiration from Kaua'i that has been embodied in song and story: The Water of Kāne (Ka Wai a Kāne) (partial quote of *mele*).

E u-i aku ana au i a oe,	A question I ask of you:
Aia i-hea ka Wai a Kāne?	Where is the Water of Kāne?
Aia i Kau-lana-ka-la,	Out there with the floating sun
I ka pae opua i ke kai,	Where cloud-forms rest on ocean's breast,
Ea mai ana ma Nihoa,	Uplifting their forms at Nihoa,
Ma ka mole mai o Lehua;	This side the base of Lehua;
Aia i-laila ka Wai a Kāne.	There is the Water of Kāne.[10]

"Therefore, the thought of sailing in search of Nihoa became fixed in the mind of Ka'ahumanu," wrote the native historian S. M. Kamakau.[11] "Say children, let us sail in search of Nihoa," Ka'ahumanu declared, and enlisted Captain William Sumner, a bluff Englishman, to relocate the island.[12] Apparently, Queen Ka'ahumanu and her husband, King Kaumuali'i, chief of Kaua'i, sailed with Sumner and landed here in 1822. The discovery of the ancient Nihoan culture established a link between prehistoric and contemporary Hawai'i.

King Kamehameha IV (Alexander Liholiho) visited Nihoa in 1857 and formally annexed Nihoa for the Hawaiian Kingdom. An excerpt from the *Manuokawai* log is instructive: "At 10 a.m. went ashore (got upset in landing). The King and Governor [Kekūanaō'a] landed at the same time in a canoe . . . a small drain of water was found near the landing but I believe it impregnated with sulfur. About a dozen seal were found on the beach and the King shot several of them."[13]

The last monarch

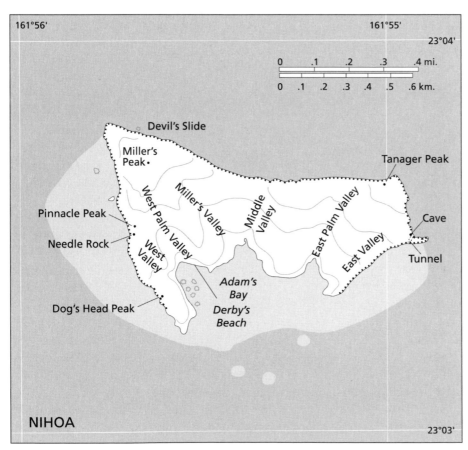

Map of Nihoa.

Members of the Liliʻuokalani entourage on Nihoa, 1885. (Bishop Museum photo)

in on a breaker, the Princess assisted by Mr. C. B. Wilson, sprang on board, the boat being instantly carried out to sea again on the retreating wave. It is safe to say that none but Hawaiian females could possibly have landed on the island and gained the boats again."[15] Included in the group was Sanford B. Dole, sent to study birds. He later participated in the 1896 revolution that dethroned Queen Liliʻuokalani and commandeered the Kingdom of Hawaii.

The cultural artifacts of Nihoa were formally investigated by Kenneth P. Emory, of the Bernice P. Bishop Museum in Honolulu. In 1923, the minesweeper USS *Tanager*, under the command of Captain Samuel Wilder King, set off with Dr. Emory and a dozen other scientists to survey the Northwestern Hawaiian Islands and record archeological, biological, and meteorological observations. (Captain King, incidentally, later became governor of the Territory of Hawaii, a delegate to Congress, and a representative to the State Legislature.)[16]

The Tanager Expedition found the extensive ruins on Nihoa and nearby Necker Island to be unlike any known from the main Hawaiian Islands. The 27-year-old Emory was placed in charge of the excavations. In pondering the cultural artifacts, he posited, "Do the ruins on these islands illustrate the enterprise of native fishermen from Kauai or Niihau, or are they the result of a resident population, or of bands of migrants? Whatever the traces are, they are sure to be faint and easily overlooked."[17]

By 1924, the *Tanager* crew was experienced in landing wooden whaleboats on rocky shelves, and they set Emory and his team safely ashore. Quickly surveying the island, Emory was immedi-

to visit Nihoa—indeed, the last Hawaiian of the royal line to reign—was Her Royal Highness Princess Liliʻuokalani, who landed on Nihoa with her entourage on 22 July 1885. "The Princess and her train had landed and visited the palms, and were returning to the shore. The island had been ransacked for birds, skins, eggs and feathers. Over 200 people had landed and worked their sweet will. . . . Doubtless there had been lunching and a good time."[14] An accidentally set brush fire swept the tinder-dry island, forcing the visitors to flee. Although Liliʻuokalani and her guests hastened to leave, they were delayed by the rising tide and seas. Two boats were swamped, causing the loss of photographic negatives and artifacts removed from Nihoa's ancient habitation sites. "If the party had been all foreigners some would have drowned," reported Honolulu's newspaper, *The Pacific Commercial Advertiser*, on 28 July 1885. It went on to say: "H.R.H. Princess Liliuokalani prepared herself for springing into the boat by assuming her bathing dress, and as the boat was swept

Archaeological excavation of large grotto, East Palm Valley, Nihoa, 1924, Tanager Expedition. (Photo by K. Emory, Bishop Museum)

East Palm Valley with habitation sites and Nihoa Palms.

ately impressed with the extent of terracing and amount of soil on this relatively steep island. All the gentler slopes were stepped with level terraces and foundations of house ruins, and the island summits were crowned with platforms of coral rubble. In several days, Emory was convinced that he had recorded all the major sites and collected all the important artifacts: stone bowls, jars, adzes, needles, and a fishhook carved from human bone in a typical Hawaiian style.

In his 1928 monograph on the archaeology of Nihoa and Necker Islands, Emory described sixty-six archeological sites on Nihoa, including fifteen religious platforms containing stone uprights along each platform's periphery. (A Bishop Museum expedition in 1984 identified another twenty-two sites.)[18] Based on the number of rock shelters and house sites, Emory estimated that 100 people could have survived long-term. He calculated that 12 out of 156 acres or 7.7 percent of the island had been terraced to grow dry-land crops. Therefore, Nihoa could have yielded up to 48 tons of sweet potatoes annually, or about 3 pounds of sweet potatoes per person per day. Combined with abundant fish and seafowl, this diet could have sustained the colonists. Yet, only six partial skeletons of adult males, females, and infants were discovered in sea caves on the flanks of Nihoa. Today, it is difficult to imagine that more than a few could thrive because of the island's shortage of potable water. There are three springs that provide several gallons a day, but the water is fouled with guano.

References

1. Clapp et al. (1977).
2. Pukui and Elbert (1965).
3. Cleghorn (1988).
4. Clapp et al. (1977).
5. Evering (1980).
6. Vancouver (1798:81–82).

Dry-land taro garden terracing on Nihoa.

7. Meares in Emory (1928:8).
8. Evering (1980:4).
9. Kamakau in Emory (1928:8).
10. Emerson (1909:257–259).
11. Kamakau (1961:9).
12. Kamakau in Emory (1928:9).

13. Paty in Emory (1928:9).
14. Bishop in Emory (1928:10).
15. *Pacific Commercial Advertiser* (1885:3).
16. Olson (1996).
17. Krauss (1988:86).
18. Cleghorn (1988).

2

Life on Nihoa

He pu'u kolo i Nihoa.

"Crawling up the cliffs of Nihoa."
Said of one who perseveres.

Nihoa seen at sunset from the northwest.

Nihoa was on the horizon, but furious 35-knot winds and 10-foot swells made landing impossible. It was 19 January 1981, and the research vessel *Townsend Cromwell* of the National Oceanic and Atmospheric Administration (NOAA) was in those scudding seas to study lobsters and deep-sea fish, survey the seafloor for deposits of manganese nodules, and drop us off at Nihoa. It was the first major winter expedition, and I was there to study seabirds while Sheila Conant, a professor at the University of Hawai'i, attended to the land birds. The pint-sized biologist has an amazing amount of energy plus a lifetime of experience with Hawaiian forest birds. Studying the unique Nihoa Millerbirds and Finches, she has spent more time scrambling up the brushy slopes of Nihoa (ever careful to avoid seabird nests) than any of her contemporaries.

Winter expeditions to Nihoa are rare because the seas are typically high and landing there is always dicey. Gelb, the dour captain of the *Cromwell*, decided it was too rough and ran past the island. We would have a second chance to visit Nihoa on the return leg of our voyage. Nihoa fell away astern as we headed northwest, entering a wilderness that is primordial Hawai'i, isolated by geologic time, continental drift, and nautical mile after nautical mile of ultramarine. I fortified myself with seasick medication for the heaping seas that we encountered upon leaving the lee of Nihoa. As we plowed through the open ocean in our steel ship reeking of diesel, my stomach churned and my head buzzed with motor noise. I scanned the radar and satellite fixes to see where the weather was coming from in relationship to where we were going, hoping for smoother seas ahead.

Southern slopes of Nihoa, looking east toward Tanager Peak.

hours of dream-state wakefulness. Wedged in my rack to keep from being tossed out, I watched thoughts, afterthoughts, reveries, fantasies, and memories slip by like a slide show. I dreamed about Sherman, my childhood buddy, waxing "the black shark," his 1963 Pontiac convertible—it's amazing what the mind retains and the stomach disgorges.

A week later, the sea was flat calm when we sidled up to Nihoa, on what was the most beautiful day of the entire cruise. We were especially grateful for this last-chance landing. Nihoa or bust! Taking full advantage of the calm before the next storm, the *Cromwell* crew hustled us into rubber Zodiacs and hauled our gear ashore with enough food and water for a month's stay. There is no such thing as a carefree landing at Nihoa, but NOAA's skilled Native Hawaiian boat handlers got us ashore. We somehow managed to catch all the gear they threw at us, but fielding a 5-gallon water jug while knee-deep in wave surge on a slick rock shelf is no easy feat. The crew left us with a two-way radio and two quarts of Neapolitan ice cream. As the ship headed away to catch lobsters, we attacked the ice cream with the swarms of flies that materialized to join us in eating the melting mess.

Nihoa has the most nearly intact coastal ecosystem left in Hawai'i because the island was spared exploitation by guano miners in the 1880s.[1] Although seabirds have left their

Counting flying seabirds while nauseous is futile, especially when birds are few and far between. I staggered to my dark bunk for many long

Nihoa Finch, male.

droppings for eons, periodic rain squalls scour the steep slopes and flush the dissolved fertilizer into the sea. Originally known as Moku Manu, or Bird Island, the description of Nihoa on the first nautical charts attracted the miners. But dismayed by the lack of guano, the miners passed on. Thus Nihoa avoided the invasion of many alien weeds and pests that plague the main islands and some of the northwestern isles. Nihoa has only five alien plant species and several introduced insects, such as ice cream flies.

Scientists have finally learned that even the most benign human presence can affect isolated island ecosystems in unpredictable ways. As with the Apollo moon landings where everything was sterilized to avoid contaminating the lunar environment, we took precautions to avoid introducing pests to Nihoa. We examined the seams of every tent and the instep of our shoes and gleaned every seed and insect from equipment before we took it into the field. We froze our food and gear, packing it in plastic buckets. It's easy for ants and

cockroaches, even geckos, to hide in cardboard boxes and paper bags. We now know too well how introduced organisms and foreign cultures can invade, change, and destroy native life-forms and ecosystems.

Nihoa provides abundant habitat and elevation zones for crevice-, ground-, and tree-nesting organisms, especially seabirds. Nearly half a million seabirds of seventeen species breed on Nihoa Island alone, including the largest Hawaiian colonies of several seabirds: Great Frigatebirds, Brown Boobies, Red-footed Boobies, and Brown Noddies. Over 90 percent of the Bulwer's Petrels, Tristram's Storm-Petrels, and Blue-gray Noddies nesting in the Hawaiian Archipelago occur on Nihoa.[2] The island also has two land bird species, the Nihoa Finch and the Nihoa Millerbird, that are found nowhere else on earth.

During our stay, Dr. Conant taught me much about "her birds" in this biological paradise, and I came away with new perspectives on evolution.

Nihoa Millerbird perches on a goosefoot (Chenopodium *sp.*) *branch.*

The surviving Northwestern Hawaiian Island finches offer clues as to how the ancestral finches evolved into new forms. Biologists believe that one lost flock of finchlike birds reached Hawai'i, successfully established themselves, and evolved into at least fifty different species called the Hawaiian honeycreepers, which comprise the majority of the forest birds native to Hawai'i. It appears from ancient bird skeletons recovered from lava tubes in the main islands, which usually preserve the bones in dry conditions, that Nihoa Finches occurred on Moloka'i in prehistoric times, and a similar species, Laysan Finches, also used to exist on O'ahu and Moloka'i.[3]

Conant's research, which began in 1980, has demonstrated that unique adaptations to the local environments occur relatively quickly. The type of food available to Nihoa and Laysan Finches on their respective islands seems to be a determining factor in the size of their parrotlike bills. Though their home islands vary greatly in elevation and microclimates, both of these finches are able to eat a variety of foods, including tough spiny seeds, fleshy plant parts, insects, and carrion, as well as seabird eggs. Conant found that Nihoa Finches have slightly smaller beaks than Laysan Finches, and though they eat similar kinds of food, the proportions are different.

She was able to take advantage of a unique situation to confirm her hypothesis of speedy genetic change in bill sizes. As insurance against potential ecological catastrophes, the U.S. Fish and Wildlife Service (USFWS) transplanted finches to other Northwestern Hawaiian Islands in the 1960s. Nihoa Finches were taken to Tern Island at French Frigate Shoals, but they did not establish a viable population. However, in 1967, the USFWS took 108 Laysan Finches to Pearl and Hermes Reef, and today about five hundred of these birds occur on North, Seal-Kittery, Southeast, and Grass Islands at the atoll.

Conant determined that in the relatively short period of time the finches were on Pearl

and Hermes Reef their bill size diverged from those on Laysan. The primary finch food at Pearl and Hermes is the tough seed of *Tribulus* or Puncture Vine. Differences in diet and natural selection of genetic variants have produced finch populations with slightly different bill sizes—even among the four islets of Pearl and Hermes Reef. (Strangely, females on Pearl and Hermes Reef are much more variable in bill size than females on Laysan.) Conant's observation that bill size can change in evolutionary "quick time" (less than 20 years!) is profound. Her study suggests how the spectacular hooked beaks of other Hawaiian honeycreepers could have evolved to allow foraging on insects,

Nihoa Palms with Pinnacle Peak in background.

as well as on seeds and nectar of the flowers of Hawai'i.[4]

The number of Nihoa Finches fluctuates between two thousand and four thousand birds. When it was discovered in 1916, it was thought to be the last undiscovered Hawaiian bird species. Its scientific name, *Telespiza ultima,* bears this out. However, in 1923, Alexander Wetmore, the biologist in charge of the Tanager Expedition, discovered the Nihoa Millerbird, *Acrocephalus familiaris kingi,* and named it for Captain Samuel Wilder King, who commanded the *Tanager.* Numbering perhaps three hundred, the pearl gray–brown millerbirds evaded detection until that late date by skulking in the brush. Especially rueful was Harold C. Palmer, the bird collector for the British financier Lord Walter Rothschild, who had been forced to bypass Nihoa in 1891 because of rough weather and missed discovering this species.

I stayed on Nihoa long enough to get to know some of its secrets, but not long enough to experience loneliness. Every day there was a new place to investigate and another facet of nature to discover. As I censused Gray-backed Tern colonies, some Nihoa Finches followed me and when some terns flushed from their nests, the finches hopped up to the eggs and cracked them with their heavy beaks. I had to leave the area immediately to permit the other terns to defend themselves against these egg robbers. Climbing toward Miller's Peak, I saw groves of Nihoa Palms, perhaps five hundred altogether on the island, growing in East and West Palm Valleys. These are truly remnants of a "lost world." Recently, botanists have discovered that up to half of the pollen found in soil cores from lowland archaeological sites in the main Hawaiian Islands comes from palms like these. *Loulu,* or fan palms, were once common plants in Hawai'i and coexisted with humans in the main islands until after A.D. 1000, when the human population expanded. People cut palms as they used more land in agriculture and they also used the 30-foot-tall palms for timber, thatch, and firewood. The wood is also reputed to have been used for spears. Polynesian Rats introduced by the early colonizers also attacked the seeds. As a consequence, the coastal fan palm forests virtually disappeared from the main islands and fan palms survive in numbers only at Nihoa and on offshore islets at Moloka'i.[5]

Nihoa Palms were probably used extensively by the ancient residents. The long-term reduction in trees could have caused a crisis for supply materials as well as increased erosion and contamination of the water supply with guano. It is possible that several drought years, during what is now called an El Niño event (when the ocean

Nihoa Palms, Pritchardia remota.

temperature warms and rain fails to fall), caused sweet potato crops to fail and underground water reserves to dry up. In addition, if the seabirds skipped a breeding season or two, Nihoans would have found their gods unresponsive and their petitions for more food and water unheeded. Eventually, they might have considered themselves utterly forsaken and abandoned Nihoa. Although this may be wild speculation, it would not be the first time a similar scenario has occurred in Polynesia. On Easter Island, located at the southeastern limit of Polynesia, archaeologists have documented an ecological disaster ensuing after all the palm trees were cut.

As I climbed the scree slopes above West Palm Valley and moved onto Dog's Head Peak, I could see other species of Nihoa plants that are also in danger of extinction. Looking more or less like a garden-variety weed, the tall, rangy amaranth is the rarest, and the last twenty-five specimens cling to the rocky slopes. The bluish succulent *Schiedea verticillata* looks more like what an endangered species "should" look like. Its fleshy, triangular leaves resemble those of no other plant on the island. This member of the carnation family numbers in the low hundreds. But my favorite plant on Nihoa is *'ohai,* a silvery leaved legume with coral-colored flowers now very rare in the main islands. The greatest population of this legume is on Nihoa where it is safe, unlike the coastal O'ahu populations that are subject to damage by insects, rats, and off-road vehicles.

At Dog's Head Peak (358 feet), I found a platform marked with coral heads, suggesting that this was an ancient fishing shrine. Higher on the ridge, the crooked finger called Pinnacle Peak beckoned me. Standing 626 feet above the sea, it is said in legend to mark the watery grave of a princess. Indeed, the demise of the Nihoa cultural connection to Kaua'i is foreshadowed in her story. A beautiful daughter of a Nihoa fisherman was desired by a prince from Kaua'i. When he came to Nihoa

Schiedea verticillata, *an endangered species of the carnation family on Nihoa.*

in search of her, she ran up West Palm Valley to escape. He pursued her until she came to the brink of the sheer sea cliff. "If you touch me, I shall jump," she cried, but still the prince reached out for her. In a moment she was gone, and the prince was turned into stone.[6] Another interpretation of the land form is more graphic. The rounded and erect rock formation was named Ka Ule O Nānāhoa by the Tanager Expedition, translated as "the male genital in a marital position"—a monument to attempted rape.[7] Finally, the geologic explanation: Pinnacle Peak is a volcanic dike created when eruptions forced dense lava through cracks in the original lava cone. As the older, softer rock eroded, the harder material was exposed and shaped by the weather into the odd formation.

I moved carefully along the perpendicular

precipice to the summit of Miller Peak where I could look straight down 900 feet to the roiling surf below. This is the edge of the world, where spirits can soar into a multihued blue realm: deep indigo seas and azure skies flared with the rag-tag flight of White Terns mating while sailing on updrafts. As I gazed on, clouds streamed by, but one puff moored in the lee of the island, something to hang my thoughts on. The late afternoon light illuminated the mint green faces of the Brown Boobies flying overhead. They swept back their wings and dangled their webbed feet, stalling in flight to get a better look at me. A curious Bristle-thighed Curlew whistled *to-wee-oweee* and I answered; it called again and I whistled it down to earth.

Below Miller Peak, I found the Albatross Plateau and censused the Black-footed Albatross colony. About fifty pairs nested here, the only flat place on Nihoa. Between the plateau and the peak is a deep cleft called Devil's Slide. I descended halfway into this short hanging valley bounded by tall volcanic dikes before I scared myself. One false step in this chute and you can free-fall to oblivion, but what a way to go!

Devil's Slide is a microcosm unlike anything elsewhere in the Northwestern Hawaiian Islands and maybe even in the main islands. In the narrow cracks that hold moisture, rare ferns grow and the largest cricket in Hawai'i lives under the moss-slick rocks. Sheila Conant discovered the cricket feeding one night on a dead Red-footed Booby. Also caught scuttling away from the carcass was a giant earwig, measuring about an inch and a half including its evil-looking pinchers for doing the devil's work. There is an ecological "guild" of giant crickets and earwigs that depends on seabirds for food. They, in turn, provide food for pseudoscorpions, wolf spiders, and centipedes.[8]

Gigantism is a phenomenon found in many animal and plant species on small isolated islands where their populations are relatively free from competition and predation. Without vertebrate predators, insects are able to take advantage of ecologically vacant niches. Some selectively lose the power of flight and develop larger size, perhaps to become bigger than other insect competitors and predators. The evolutionary process that yields giant crickets and earwigs in Hawai'i has occurred on other island groups. The Weta, a New Zealand cricket, has evolved into one of the heaviest insects in the world. Weighing several ounces, the Weta is an extreme example of island gigantism and is the ecological equivalent of a field mouse. The several species of New Zealand Wetas are relics of the Carboniferous Period, about 300 million years ago, when giant insects ruled the world.

At least 485 species of insects and spiders occur in the Northwestern Hawaiian Islands. Of these, about 300 have been accidentally introduced; 100 are indigenous, occurring naturally in Hawai'i as well as other Pacific islands; and 80 are endemic, unique to specific islands.[9] At least 40 of the 187 terrestrial arthropod species (insects, spiders, and crustaceans) on Nihoa are endemic (found nowhere else in the world).[10]

A passing squall momentarily interrupted my big-bug reveries and I slipped on a wet tussock that was home to "coneheads"—not the well-known TV characters from *Saturday Night Live,* but a species of katydid, or grasshopperlike insect. Most species in Hawai'i inhabit moist mountain forests, but the largest in all the archipelago lives in the bunchgrass of Nihoa. They may have existed in grasses on the main islands at one time, but introduced predators, notably Polynesian Rats and predacious Argentine Ants, have probably extirpated the katydids, and introduced herbivores and human-induced fires eradicated native grasses. A major difference separates the Nihoa and main island species. The impetuous males of the main islands jump on the females to mate, sometimes getting killed in the process. The Nihoa males do not pounce but sing until the female facilitates mating.[11]

common on many Northwestern Hawaiian Islands and can easily be accidentally transported to Nihoa by careless visitors.

References

1. Gagné and Conant (1983).
2. Harrison (1990).
3. Olson and Ziegler (1995).
4. Fleischer et al. (1991).
5. TenBruggencate (1994).
6. Wetmore (1925).
7. Olson (1996).
8. Gagné and Conant (1983).
9. Conant et al. (1984).
10. Strazanac (1992).
11. Strazanac (1992).
12. Conant et al. (1984).

Sheila Conant scrambling up Nihoa's slopes with insect net.

Land snails are another group of animals that once occurred throughout the main Hawaiian Islands until introduced predators almost wiped them out. The majority of the coiled mollusk species are small and obscure, living in dun-colored shells about half an inch across. More than two hundred species are extinct, but at least six species still live in bunchgrass on Nihoa, a relict community of what once occurred in the main islands.[12] Threats from introduced insects, whether predators or competitors; rodents; weed seeds; disease; and fire are ever present and pose a substantial risk to these vulnerable invertebrate populations. Ants, flies, cockroaches, and numerous weeds are already

3

Cromwell, Do You Copy? Over

Nihoa i ka moku manu.

Nihoa, island of birds.

Derby's Beach, Nihoa.

Every few days on Nihoa, we were reminded that it was still winter. After a midnight rainstorm scoured our camp, I spent the next morning recovering, wringing out, and cleaning up. The pallid, gray morning melted into a clear, windless afternoon. The calms before and after the storms allowed us to regroup before the next front hit. In the late afternoon, I moved down to the tide pools and watched the black rock crabs (*'a'ama*) skitter away and horned blennies with white clown lips flip from one tide pool to another. The waves surged over the rocks splotched with coralline algae the color of calamine. The color was all too familiar because I was spotted with the same lotion soothing my tick bites. Also clinging to the rocks were *'opihi*, giant limpets wearing chinaman's hats, and *hā'uke'uke*, purple sea urchins with blunt spines radiating out from the center.

I stared out at the brilliantly blue water ever in motion and noticed white water where there weren't any reefs. A pair of courting Humpback Whales floated belly up, waving their huge pectoral fins, slapping their flukes, and frothing the ocean. These leviathans migrate from the higher-latitude North Pacific summer feeding grounds to their winter breeding/calving areas around the main Hawaiian Islands. Their numbers peak around one thousand in late January through February and remain fairly constant through mid-March. In April, they begin migrating out of Hawaiian waters and by late May or early June, the last whales have departed for Alaska. During the later stages of winter migration, humpbacks are occasionally sighted around Nihoa and French Frigate Shoals.

On my way back to camp, I paused to wash up at the water seep, a smelly puddle of *"eau de guano."* Nearby, a black plastic shower bag hanging suspended from a rock provided a quick rinse off with solar-heated fresh water. When I reached for the soap lying next to the pool of acrid water, I noticed a cricket floating in the water. I imagined him singing, "I'm no fool, no siree, I'm going to live to be a hundred and three." Not! I hummed a few bars, gleefully aware that Conant would be elated when I showed her "Jiminy." She was, indeed, because it appears that this was an adult of a species of flightless cricket unknown to science. However, nymphs, a developmental stage of crickets and some other insects, were necessary for conclusive species identification. So that night Conant placed baited tin cans just above the high-water mark and caught several. Upon her return to Honolulu, she provided entomologists at the Bishop Museum with a peek at the nymphs of these flightless, algae-eating crickets. Imagine crickets living in the shoreline splash zones. How Hawaiian! They probably surf—or at least disperse between islands on floating debris.[1]

Tristram's Storm-Petrel, Oceanodroma tristrami.

Biological surprises seemed to be around every corner on Nihoa in winter. Returning from Tanager Peak (852 feet) one day, I stepped into a depression and inadvertently collapsed a bird burrow. When I dug it out, I discovered the first live Tristram's Storm-Petrel chick ever recorded on Nihoa. This rare nocturnal seabird nests on only a few islands in Hawai'i and Japan, so that night we went hunting for "stormies." One of my favorite type of birds, storm-petrels are the smallest of seabirds and were called Mother Carey's chickens by the seamen of yore. Mother Carey is a corruption of Mater Cara, Italian for Blessed Virgin Mary, Queen of Heaven. Sailors predicted that a storm was approaching when storm-petrels appeared. In calm weather, the birds may patter their webbed feet on the sea surface, harkening another religious association, with St. Peter, whose name was the basis for the appellation "petrel."

I guesstimated that about two thousand "stormies" lived on the island. I loved handling them for the musky smell they emit—admittedly an acquired taste. We captured, banded, and released ten, and while we worked, their tiny voices burped under the rocks. In the night chorus, they joined the tens of thousands of Bulwer's Petrels similarly calling from their hidden underground nests. With them, the eerie caterwauling of Wedge-tailed Shearwaters gave voice to the spirits that inhabit Nihoa.

Contending with the island's steep terrain in the ever-present wind was a challenging but small price to pay for the opportunity to conduct novel research. By day, I worked on the bird cliffs alone and without ropes.

Blue-gray Noddy on Nihoa.

This precarious area had a dizzying view of Derby's Beach 100 feet below. If I fell, I hoped that I might land on a fat monk seal. It was worth the risk, for I had a unique opportunity to study the breeding biology of the Blue-gray Noddies, which only breed in the winter. George C. Munro wrote: "This neat little bird is grey all over, with some little differences in shade and very little white on the wings. . . . It is familiar yet wary. . . . We did not collect it on the Rothschild Expedition along the Hawaiian Chain in 1891."[2]

The smallest of the Hawaiian terns, Blue-gray Noddies flit above the tropical Pacific picking up Water Striders, the only insect known to inhabit the open ocean. In the winter, when the insects diminish, noddies prey on baby *mahimahi*, a fish favored by Hawaiians and tourists alike. Perhaps the seasonal abundance of its prey permits the female Blue-gray Noddy to produce an enormous egg. These small birds lay the largest egg relative to their body size of all birds, about a third of their 2-ounce body weight. Blue-gray Noddies shelter their eggs in a *puka*, a small hole originally formed by gas bubbles in the basaltic

lava. Every few days, I scrambled up the rocky ridge to check their nests in the steep cliffs. When the eggs hatched, I measured the chicks' warm, pliable beaks and legs with cold steel calipers to determine their rate of growth.[3] As I quickly measured these fluff balls, the adults would hover by, churring at me, most displeased by my disturbance. Despite the thrill of working with these lavender gray sea sprites, I was getting tired of hassling them and glad to wrap things up.

After 3 weeks, we looked forward to the arrival of the *Cromwell*, but the sea swells were beginning to build. Sure enough, our scheduled pickup was nixed—a "no-go" because of the rising surf. About every 15 minutes, a mountainous set of four or five waves would sweep into the cove,

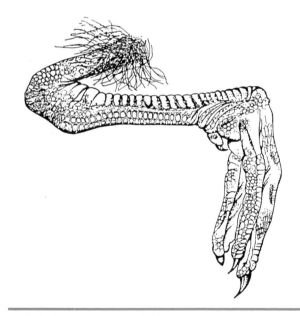

Foot of a Blue-gray noddy chick.

24

slamming against the roots of Nihoa. The surf continued to grow to 25-foot breakers of immense power, sending up an aerosol cloud of salt spray and filling Adam's Bay with sea foam. The usual northeast trades were blowing hard, but these awesome swells came from the southwest, probably generated by a tropical cyclone 3,000 miles away.

Preparing to abandon the island at a moment's notice, I hauled some heavy boxes down to a lower lava shelf. As I began to climb back to camp, a rogue wave rolled into the cove and carried off one of the boxes. I helplessly watched the box, which held our radio, float away in the frothy maelstrom until it sank—a reminder that the ocean giveth and the ocean taketh away. I could not escape the sinking feeling of losing our connection to the outside world, as well as our most expensive piece of equipment. I hit rock bottom. Luckily, although as dog tired and dirty as I, Conant was ebullient. Covered with grit and salt, we both agreed: this was the essence of life—to love nature without trusting it.

The unrelenting wind challenged everything we did. Only the tents provided escape. I hunkered down at 7 P.M., waiting for the brutal storm to pass as every gust threatened to rip open my shelter. The driving rain pushed the tarp to within inches of my face. The roaring surf accompanied the wind whistling through the ropes. Something suddenly snapped, and the tarp whipped throughout the long night. In the morning, about an inch of water had collected in the corner of the tent, and my clothes and sleeping bag were soaked.

Winter storm waves hit Nihoa's cliffs.

Fortunately, the dawn also brought the eye of the storm and a chance to escape. When we saw the *Cromwell* on the horizon, we frantically collapsed our wet tents and shuttled the last of our gear to the surf ledge out of wave reach. As we waited for their tentative approach, I recalled how some early explorers like Captain "Mad Jack" Percival of the schooner USS *Dolphin* had found it more difficult to depart than to land. He was stranded on Nihoa for 2 days, and his wooden whaleboat was swamped repeatedly in the breakers at Derby's Beach. Eventually, the boat was severed amidships and he had to be hauled through the surf by a rescue party with a rope tied around his waist.[4] And Alexander Wetmore, leader of the Tanager Expedition, had been swept off the landing ledge by a tremendous wave and dumped into the drink. He was plucked from the sea foam by his hair and fingertips.[5]

The *Cromwell*'s rubber boat cautiously approached the lava ledge as the waves surged up and down. When a set of breakers subsided, we pitched our plastic buckets into the Zodiac, which quickly backed up and transferred the gear to the Boston Whaler standing by in deeper water. Finally, everything was off the island except for us. Getting in the boat was equivalent to jumping in an elevator as it speeds by. A misstep could be disastrous. We watched the waves closely to time our jump and took turns leaping into the boat on a descending wave (better than meeting the deck on an ascending thrust). The expert Hawaiian boatman threw the motor into reverse and the boat revved away from the rocks. We sped out to the Whaler where we transferred our personal gear and ourselves for the ride back to the *Cromwell*. We again timed our jump onto the ship's Jacob's ladder and clambered up while the ship heaved like a whale. Soon all the boats were hoisted aboard and the *Cromwell* weighed anchor and headed for Kaua'i. I had lost a lot of weight as well as my sea legs and hoped to make it up on the ship's

Nihoa in perspective.

chow, but I was immediately seasick and could not stomach food until we were in the harbor. I laid in my bunk reeling from the departure experience, feeling both lucky to be in one piece and privileged at having a unique winter perspective of Nihoa.

References

1. Howarth and Montgomery (1998).
2. Munro (1944:62–63).
3. Rauzon et al. (1984).
4. Clapp et al. (1977).
5. Wilson (1923).

4

Moku Manamana

Ua hala nā kūpuna, a he ʻike kōliʻuliʻu wale nō kō keia lā, i nā mea i ke au i hope lilo, iō kikilo.

"The ancestors have passed on; today's people see but dimly times long gone and far behind."
Pukui et al., *Place Names of Hawaii*

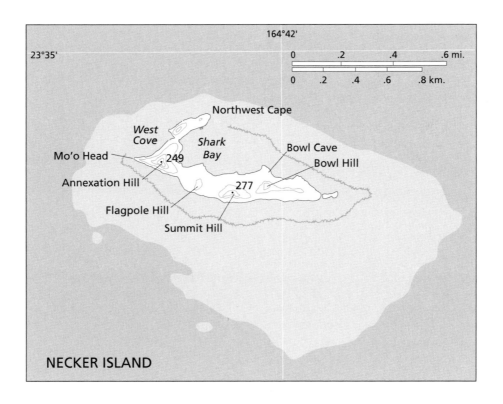

Map of Necker Island.

About 155 miles northwest of Nihoa lies Necker Island, a precipitous ridge roughly 40 acres in extent and, at its highest, 277 feet above the surf. It is all that remains of a submerged volcano, over 10 million years old, that was once comparable in size to Oʻahu.[1] The island forms a hook shape that encloses Shark Bay. The Hawaiian name Moku Manamana means "branching, or pinnacled islands" and accurately describes Necker.[2] Historically, the name Moku Manamana is known from ancient chants but was not known to be directly associated with Necker. However, because Necker is the most obvious "branched island," ancient Hawaiians did live there, and the name is from antiquity, it seems reasonable to conclude that the name Moku Manamana referred to Necker Island.

It was the peripatetic captain Jean François de Galaup, Compte de La Pérouse, who rediscovered and named Necker Island on 4 November 1786. He wrote in the ship's log: "It does not exhibit a single tree, but there is a great deal of grass near the summit. The naked rock is covered with the dung of birds. . . . The banks were perpendicular, like a wall, and the sea broke so violently against them, that it was impossible to land. . . . I called it Ile Necker [in honor of Monsieur Jacques Necker, French Minister of Finance under Louis XVI]."[3]

Also unable to land was Captain William Paty, who was sent by the Kingdom of Hawaii in 1857 to explore Necker and other northwestern islands and claim them for His Majesty King Kamehameha IV. As late as 1890, it was still unclear who held title to Necker Island when the British government moved to acquire it for a cable station. An underwater cable had been proposed to link Australia with Canada and the rest of the world's telegraphic network. When the British warship *Champion* showed up in Honolulu Harbor, the Hawaiian government immediately dispatched the minister of the interior, Captain James A. King (father of Samuel Wilder King), to formally claim the island. Led by Captain Freeman of the steamer *Iwalani,* the party landed after considerable difficulty on 27 May 1894 and raised the flag in the

Annexation Party on Necker Island, 27 May 1894. Left to right: *First Officer William Gregory, Minister James King reading Proclamation of Annexation, Captain Freeman, and three sailors. Photo by Ben H. Norton, chief engineer of the* Iwalani. (Bishop Museum photo)

name of the Kingdom of Hawaii.[4] With the island finally claimed, the British tried to negotiate for its use. Sanford Dole, president of the Provisional Government of Hawaii, was not interested in having the British do business in Hawai'i because he would run the risk of alienating the United States and its burgeoning interest in the island kingdom.

Landings became more frequent after 1900, resulting in detailed observations of the unique geology and biology. In 1902, the U.S. Fish Commission sent the *Albatross* up the northwestern island chain, and scientists discovered a new bird species. The Blue-gray Noddy was described from Necker Island in 1903 and initially called the Necker Island Tern (populations were later found on Nihoa, La Pérouse Pinnacle of French Frigate Shoals, and on Gardner Pinnacles). Nine years later, the U.S. Revenue Cutter *Thetis,* on patrol to apprehend Japanese bird poachers and to deliver mail to Midway, visited Necker Island. Landing from wooden boats on wave-washed lava was impossible in high seas, so entomologist George Willett, desperate to get ashore in 1912, stripped off his clothes, swam ashore, and surveyed the island. When the Tanager Expedition visited in 1923–1924, basic archaeological and biological investigations were completed.[5]

I first visited Necker in 1980 when the *Townsend Cromwell* dropped a team of biologists on the island for an overnight seabird survey. In contrast to Nihoa, Necker Island is barren. Only five species of flowering plants exist.[6] Yet, on the stony ground and in the scant vegetation, 60,000 seabirds of sixteen species roost and nest.[7] While counting seabirds, we searched the crevices in the stone walls looking for storm-petrels while shearwaters moaned like ghouls from their burrows. Sooty Terns rested on unexploded ordinance left over from Navy bombing practice, and boobies slept soundly perched on ancient Hawaiian shrines. After our survey of night birds was complete, we looked for a place to sleep. The only level place to camp was on the central stone altar where I spied the laser red eyeshine of wolf spiders on the prowl. Before we laid our sleeping bags down, we blessed the site with an offering, a shot of Johnnie Walker Red, hoping that the gods of Necker would be appeased and allow us a restful night. As we crawled into our bags and pulled a tarp over us, it began to rain. It was a wet night, with sleep robbed by wailing shearwaters and biting bird ticks, proving that

The first Bird Reservation Warder, Gerrit P. Wilder, and party on Necker Island, 1919. (Bishop Museum photo)

Necker Island cliffs with Shark Bay and Northwest Cape in background.

whiskey doesn't work. We were not welcome on this island of sacred stones.

Back in Honolulu a few weeks later, I was anxious to share my impressions with the man who pioneered Hawaiian archaeology. I met Kenneth P. Emory at the beginning of my career and the end of his. When I visited him at the Bishop Museum, full of questions, he gave me his monograph to read, signed with his spidery, blue-veined hand, "With best wishes, Keneti." He had explored Nihoa and Necker Islands very early in his career, one that eventually took him to every corner of the Pacific and earned him the honor of being called simply Keneti by many. Of all the places he had seen, it was Necker Island that intrigued him most.

When Emory first approached Necker in 1924, he noticed that the profile of the island looked like a *mo'o*, a spiny lizard-god or eel-god. When he landed on the island, he was surprised to find fifty-two archaeological sites, of which thirty-three were religious temples. Emory determined that the spikes he had noticed in the island's profile were slivers of basalt rock pointing skyward from the stone altars. He suggested that these uprights were the "backrests of the gods," seats of power at a shrine to the ancestral gods. Emory recognized that these religious sites were fundamentally different from other Hawaiian sites, so he designated them *marae*, the Tahitian word for temple, rather than *heiau*, the Hawaiian name.[8] Emory's hunch was confirmed the following year when he was exploring the remote interior of the island of Tahiti.

Necker Island crouched in the sea. "May we not suppose that here flourished an ancient religion ... whose rites were so fearsome or so holy to the devotees as to preclude their celebration in public, or their mention in story, so that with the decadence of the older people all knowledge of their form has disappeared?" (Wetmore 1925:79).

While clearing the brush away from a small family shrine, three upright stones caught his eye. These isolated valley sites, 2,800 miles south of Hawai'i, provided a link connecting the early Tahitian and early Hawaiian cultures, confirmed because the Tahitian and the Necker sites were both untouched by "progress."[9]

In the Tuamotu Archipelago, east of Tahiti, Emory met an old man who had worshiped at his family shrine in his youth, before missionaries deposed his gods. The elder informed Emory that the odd number of backrests allowed the main god to be in the center, surrounded by lesser deities also important to family worship and individual guidance. Through the living memory of this one man, Emory glimpsed the old-time religion of the Polynesians and clinched his belief that cultural elements at Nihoa and Necker Islands

are characteristic of ancestral Polynesian society—a culture that existed before the thirteenth century in Hawai'i. He wrote, "The similarity of the Necker sites to those of Tahiti and the Tuamotus does not just *suggest* a link but *is* a firm link!"[10] Ten years later Emory concluded that the monoliths of Easter Island at the extreme southeastern corner of Polynesia are derived from the same original architecture taken to the extreme—massive representations of the gods themselves.[11]

As at Easter Island, the *mana*, or religious energy, of Necker Island is best expressed in unique stone images. Emory led me from his basement office in the Bishop Museum to a display in the main hall. In the showcase were four stone figures, 8 to 18 inches tall, carved in a design and manner reminiscent of statues found in the Marquesas Islands. Emory referred to Necker as "Pecker Island," basing his nickname on the conspicuous sex of the stone men as well as the manner of their manufacturing. Only ten of these moon-faced stone carvings exist at museums in Hawai'i and Germany.

The first Western sailor to find the images

Necker Island stone images. (Bishop Museum photo)

in 1879 wanted to bring them on board, but the Hawaiian sailors protested and said it would surely bring death if they were removed.[12] The figures were collected by the Annexation Party in 1894. Seven of these images were taken from one *marae,* the central religious site. The log book of the *Iwalani* reads: "Captain Freeman found several old images and idols in a good state of preservation, except for the injuries received by exposure to the weather. A number of these idols were brought back by us as curios."[13] "The idols were brought to Honolulu, set up and placed on exhibition in the windows of the Golden Rule Bazaar, where they may be seen today. How long these heathen gods have been on Necker Island, or how long the barren rock was inhabited by man, are problems which must remain unsolved."[14]

Other intriguing artifacts were discovered on Necker Island as well. A stone dumbbell was found whose "ends were only roughly hewn but the handle was undoubtedly carved down. The Hawaiians said that when sharks were lured to the shallow places that this dumbbell was used to kill them by beating them on the head. I brought the dumbbell with me and left it in San Francisco."[15] The Annexation Party log reads: "One great curiosity that we found looked like a piece of stone, but, on close inspection, it was thought to be petrified flesh. It was on a stone altar, and must have been the offering to one of the ancient gods."[16] Perhaps the Necker Islanders left shark meat on the altars to petition the gods for sustenance and rescue.

It seems most likely that Necker was used for religious purposes during short-term visits rather than long-term habitation.[17] The Necker visitors lived in nine rock shelters and caves that could have sheltered up to two dozen people for a short time. In the harsh environment of Necker, drinking water was surely the limiting factor. One of the largest caves was a few feet from two freshwater seeps known on the island. Emory thought it would be possible to collect 5 gallons of water a day from this source, which was highly contaminated with acrid salts, presumably leached from bird droppings.[18]

Bowl Cave, the largest shelter, yielded rare artifacts: a boat-shaped bowl, seven adzes, a chisel and hammerstone, three sinkers, one awl, and a fragment of *wiliwili,* the lightest wood in Hawai'i, commonly used for outriggers on canoes and net floats. Emory also discovered a T-shaped object made from basalt that he noted resembled a bird-snaring perch from New Zealand. If the Necker object was a bird snare, it lacked an obvious place to secure the snaring loop. On an island with tame birds, such as Necker Island, it would not have been necessary to snare birds. Instead, the crosslike object could have been used in religious ceremonies on altars. Indeed, the frigatebird was a "royal bird" and may have been worshipped.[19] Or perhaps the perch was for "carrier pigeons." Frigatebirds may have been used for communication among wayfinders. Birds taken from Necker as chicks and raised in another island's village would naturally "home" back to the place of their hatching. A bird could be taken on a voyage and released when a canoe reached its destination. The frigatebird would return home and indicate safe arrival—or be the last message from a canoe that never returned.

Emory dug deeper into the packed dust of Bowl Cave and found human remains of someone he estimated to be 5 feet, 3.5 inches tall. "The upper end of the right femur being blackened as if by carbonization has led to the suggestion that the bones are evidences of cannibalism. Against such an assumption is the fact that the best preserved of the bones are not broken for the extraction of marrow and the poorly preserved femur and tibia were probably broken by weathering. The long and heavy bones of the leg were most in favor among fishermen for the manufacture of fishhooks and it may be for this purpose that the bones were taken to Bowl Cave."[20]

U.S. Navy aircraft squadron approaching Necker Island: altitude 2,000 feet, 9 April 1933. (Bishop Museum photo)

In these bones I thought I glimpsed the basis of the fabled *menehune* of Hawai'i. In island tradition, the "elfin people" are the magic wee folk who completed major projects in one night. The *menehune* were an aboriginal race of Polynesians in Hawai'i. The most sophisticated architectural stonework in Hawai'i is attributed to them. Remnants are visible on Kaua'i at the Menehune Ditch, where cut stones rest flush against each other. The finely carved stone images of Necker Island may be as close as we can come to seeing the *menehune*.

The Nihoa/Necker culture may have originated in the lush and populous Nāpali coast area of Kaua'i, permeating to the outlying dry islands of Ni'ihau, Ka'ula, Lehua, and Necker. Legend holds that sailors climbed a hill on Ni'ihau called Pu'u Ka'eo and looked toward Nihoa, below the horizon 125 miles away. In fair weather, canoes departing for Nihoa would begin by sailing to Lehua, the 700-foot volcanic crescent right off Ni'ihau that serves as a geographic pointer to Nihoa.[21] People from Ni'ihau might have sailed to Nihoa on the spring winds to harvest previously planted sweet potatoes and to collect bird eggs and the red tail feathers of tropicbirds and iridescent black plumes from frigatebirds. These feathers were made into *kāhili*, plumed staffs that functioned as royal and funerary standards.[22] Perhaps newly married couples would go to Nihoa to learn to live together. If all went well they would return on the *kona* winds in the fall and a baby would be delivered in the spring.[23]

The ancient culture was probably overtaken by subsequent invasions from the Marquesas and later Tahiti-based cultures that came to dominate the main Hawaiian Islands. Legends recall that a priest named Pa'ao led a cultural infusion from Kahiki, the land of the gods, around A.D. 1000. When the aboriginal culture was overtaken by successive waves of colonists from the south, Nihoa and Necker were abandoned and survived only as remnants of oral tradition on the islands of Ni`ihau and Kaua`i.

Wainalia was the man
And Hanala`a was the woman,
Of them was born Ni`ihau, a land, an island.
A land at the roots, the stem of the land.
There were three children among them
Born in the same day,
Ni`ihau, Ka`ula, ending with Nihoa.
The mother then conceived no more.
No island appeared afterwards.[24]

Tradition also recalls that the Polynesian forefathers returned to Kahiki, by way of Nihoa. The legend from honored antiquity of Kepaka'ili-

'ula, a chief of Hawai'i whose lineage is tied to the god-chiefs, describes his voyage. Backsighting Nihoa, he set sail, driven by the trades. The canoes would soar like shearwaters, returning from the edge of the world.

When Kepaka'ili'ula departed from the shore of
      Ni'ihau,
he directed his canoe straight to the ocean voyaging
      path that is near Nihoa;
which is the limit of the little islands of Hawai'i Nei.
With the lighting of the next day, he was close to that
      island which is without people; it was then that
      he saw the birds flying about the island which is
      their residenc e ... [25]

The degree of familiarity that the ancient Hawaiians had with the other Leeward Islands will never be known, but myths and legends surrounding Pele, the fire goddess, contain terms that may refer to other islands northwest of Necker. Pele dwelled in Kahiki (Tahiti), but was forced to re-locate by a jealous elder sister. Pele and her faithful siblings set sail on a course that took them northwest of Hawai'i. "Pele . . . goes by way of Polapola, Kuaihelani (where Kāne hides the islands), and other islands inhabited by gods (Mokumanamana) to Niihau. . . ."[26] In another version of the myth, Moku Papapa, which means "low sandy island," is where Pele's brother, Kāne-milohai, built up a small island. Such a description could refer to French Frigate Shoals with the diminutive La Pérouse Pinnacle. There are other names for islands supposedly beyond Nihoa and Necker: Moku Akamohoali'i, Kānehunamoku, Hanaka'ie'ie, Hanakeaumoe, Ununui, but which island is which, no one knows, for the meanings are lost in time.[27]

Recently, the Hawaiian Lexicon Committee developed a set of names for the Northwestern Hawaiian Islands. They include the original names mentioned in legends: Mokumanamana, meaning "branched island" for Necker Island and Mokupā-papa, "flat, sand island" for French Frigate Shoals. The committee also conferred the following ap-pellations on the remaining islands: Pūhāhonu is Gardner Pinnacles, which means "surfacing of a sea turtle for air." Maro Reef is Nalukakala, which translates as "surf that arrives in combers." Laysan Island is Kauō and means "egg," representing the island's shape and the abundant seafowl. Lisianski Island is Papa'ā-poho, a flat island with a depres-sion. Pearl and Hermes Reef is Holoikauaua, a name that cele-brates the monk seals that haul out there. Midway is Pihemanu, which means "the loud din of birds." And finally the Lexicon Committee has chosen Kānemilo-ha'i to represent Kure, the last of the emergent Hawaiian Islands.[28]

Landing on Necker Island is like jumping off a moving elevator.

References

1. Armstrong (1983).
2. Pukui et al. (1974).
3. Emory (1928:53).
4. Clapp and Kridler (1977).
5. Clapp and Kridler (1977).
6. Clapp and Kridler (1977).
7. Harrison (1990).
8. Riley (1982).
9. Krauss (1988).
10. K. P. Emory, 1983, personal communication.
11. Krauss (1988).
12. Atkinson in Emory (1928).
13. Emory (1928:55).
14. *Hawaiian Star* in Emory (1928:56).
15. Atkinson in Emory (1928:54).
16. Emory (1928:55).
17. Cleghorn (1988).
18. Emory (1928).
19. Keauokalini in Rose et al. (1993).
20. Emory (1928:99).
21. Tava and Keale (1989).
22. Rose et al. (1993).
23. Tava and Keale (1989).
24. Fornander in Tabrah (1987:12).
25. *Ka Hōkū o Hawaii,* 6 May 1920, p. 1 (Kepa Maly, personal communication).
26. Rice in Beckwith (1970:169).
27. Tava and Keale (1989).
28. Kimura (1998:27).

Hōkūle'a

He noio 'a 'e 'ale no ke kai loa.

"A noddy tern that treads over the billows of the distant sea."
An expression of admiration for a person outstanding in wisdom and skill.

Hōkūle'a sails full of wind.

Nihoa and Necker Islands have awaited the return of voyaging canoes for centuries. In 1994, one was finally poised to *ka'alalo*—sail to leeward. *Hōkūle'a,* the canoe that has awakened Polynesian pride and consciousness wherever it has sailed, is a fiberglass and plywood–hulled replica of the first canoes that sailed to Hawai'i. *Hōkūle'a* is an important symbol of the Hawaiian cultural renaissance as well as a vital link to the Polynesian ancestry. The crew of the canoe began to recreate the fantastic Pacific explorations of the ancient Polynesians by voyaging from the Big Island of Hawai'i to Tahiti, the heart of Polynesia. On their return home to Hawai'i in 1976, as on the original explorations, the navigators followed the starpath north, particularly following the northern star called Arcturus, or Hōkūle'a in Hawaiian, which reaches its zenith above the Hawaiian Islands.[1] The Hawaiian Archipelago spans 1,600 miles, with islands roughly 100–200 miles apart, so the navigators were likely to make landfall somewhere. Like the ancient voyagers, the crew could have seen Nihoa—the tallest of the Northwestern Hawaiian Islands.

For over two decades, *Hōkūle'a* had sailed throughout Polynesia but had never been to Nihoa. In 1994, the crew decided to sail to this pivotal island on a training mission. The master navigator of the canoe was Nainoa Thompson, a youthful, charismatic, but quietly intense man. He was the wayfinder who helped guide *Hōkūle'a* to and from Tahiti on several voyages of reconnection. Sailing her over 50,000 nautical miles, he had become aware of every shift in the wind and what it might mean for the craft. On this voyage, he was joined by first mates Kimo Lyman, a fireman from O'ahu, and Tava Taupu, a wood-carver from the Marquesas Islands. Because I had been to Nihoa and knew the best landing sites, and because a biologist-escort was required by the U.S. Fish and Wildlife Service to make sure that no organisms were accidentally introduced to Nihoa, I was invited to join the expedition.

The canoe sailed from home port O'ahu to position herself off Kaua'i in an attempt to find the forgotten water road to Nihoa and reach the northern tip of the Polynesian triangle, which also includes New Zealand and Easter Island. Along the way, *Hōkūle'a* also lent support to the Polynesian community, as Kaua'i hosted the annual Tahiti fete. Dancers and drummers from Polynesia, as well as the mainland, came to compete for recognition and prizes and to enjoy the cultural affinities, fellowship, and food in the spirit of *aloha*.

This trip was also designed to be a training sail for the 1995 voyage of *Hōkūle'a* into the South Pacific to rendezvous with other Polynesian sailing canoes. Thompson was tutoring four young navigators, members of a new generation, in the art of celestial navigation, which he had learned at the side of virtually the only practitioner, Mau Piailug. Piailug is the last living link in the memory of a collective navigational experience of a culture perpetually in touch with the movements of the sea and stars. A native of Satawal Atoll in Micronesia, he passed on to Thompson his knowledge of wayfinding in the ocean wilderness.

To navigate across thousands of miles of empty ocean, the ancient mariners devised ways to divide the trackless horizon and a language to communicate directions. By day, navigators followed the path of the sun, watched the waves, and especially felt the subtleties of the wind. By night, they experienced every wave, every gust,

Tava Taupu on Hōkūle'a.

every screeching seabird. Over two hundred star-paths had to be memorized: their rising and setting positions throughout the night relative to the moon. Colonization of the Pacific Ocean is testimony to the Polynesians' development of the fine art of reading their watery world. Orienting by the forces of the planet—sea, swell, wind, stars—their canoes moved like a seabird gliding over the waves, wings full of wind. For the ancient argonauts, oceanic intervals of space and time were part of their daily reality. Critical observations, dreams, myths, and stars guided them to whatever might lie over the horizon. Thompson adapted the information from Micronesia and developed a "compass" to use from the Hawaiian perspective. The voyages of *Hōkūle'a* proved it possible to navigate through the Pacific Ocean according to this venerable wisdom.

In 1994, *Hōkūle'a* was ready to visit Nihoa, where the spirit of the ancestors beckoned. As the *Hōkūle'a* crew waited for the right winds, doubts danced in my mind: What the hell was I doing here? I was the scrawniest of the twelve men and two women of Thompson's brawny crew and a California *haole* (white person), but the only one who had been to Nihoa before. The crew had all learned the ropes of sailing *Hōkūle'a* and were keenly aware of the privilege of being on the venerable canoe. In spite of my weaknesses, they welcomed me into their *'ohana,* or family, for they knew from experience that the way in which a crew works together determines whether a sail is successful or not. We would all depend on each other to survive the crossing. I just hoped I was up to the rigors of the trip.

My special contribution would begin if we got to Nihoa, and it was getting chancier all the time. The all-weather radio station had forecast a tropical storm heading for Kaua'i. The last big storm in Hawai'i, Hurricane 'Iniki, had hit Kaua'i in 1991 and was still on everyone's mind. Heaps of sodden plasterboard and twisted aluminum still lying in the abandoned sugarcane fields were daily reminders. Empty storefronts and the crownless palm trees evinced the power of 'Iniki. The old Līhu'e theater, the last vestige of the days when sugarcane was king, looked beyond repair.

As we waited and watched the tropical depression approach, we were lucky to be distracted by the Tahiti fete. The presence of *Hōkūle'a* in Nā-wiliwili Harbor was a festival highlight, and the crew was especially popular. A hotel provided a free room for us, now totaling fifteen. The dank doldrums that soon set in made the accommodations smell not unlike a cramped boat deck. Coconut palms dripped throughout the windless

day as the rains came on, steadily and straight down. Some of us went to visit a family who provided local food: swordfish *poke,* Kaua'i-bred grilled steaks, sour *poi,* and sticky rice. We watched Nigeria beat Argentina in the World Cup soccer game as rain poured on the tin roof. Back at the hotel, the latest forecast confirmed the obvious—local flash flooding, but Thompson anticipated that the diminishing depression would pass by late afternoon. Sure enough, toward sunset, a crack appeared in the clouds in the eastern sky and the winds freshened, a sign that the trip still might go as planned.

Thompson called for volunteers to form the night watch on the canoe in the harbor. I jumped at the chance to have at least one night on board, just in case we didn't sail. First mate Tava also agreed to stand watch, though it required a midnight swim to the canoe moored off the pier. I stripped and slid into the dark water. Something about swimming at night makes the water feel colder than it really is. I was shivering when I climbed up the lines of the canoe. Tava, about twice my bulk, appeared unfazed. I appreciated that the Polynesians may well have been genetically selected for their aquatic environment. Survival of the fittest *and* fattest. Only those with insulating layers of fat might survive extended periods of low rations and constant wetness, especially when sailing to the edges of subtropical Polynesia.

The deck smelled of sweet apple bananas that swung in rope baskets. Tava threw his bedroll on the outdoor canvas bunk lashed to the bamboo gunwales. He showed me the bunks in the hull and also pointed out the "head" at the stern, a toilet seat suspended over the water. Expecting to awaken at 5 A.M. to cast off, I tucked into my bunk in the canoe with the high tide of the summer solstice chafing the mooring lines. As I drifted to sleep, I wondered about the Hawaiians who had sailed between Kaua'i and Nihoa. Did they sail there regularly? If we were having problems and

Kimo Lyman and navigators on Hōkūleʻa.

from the flying buttress of an outhouse. Kimo Lyman had once parted company with the boat. He was taking care of business when the lifeline around his waist snapped. Kimo was overboard in an instant, but luckily Thompson spotted his hat in the water and sensed what had happened. He threw a marker buoy over, turned the canoe about, and saved the lucky Lyman from his fate.

At sunrise we were hailed from shore and told to return to the hotel. There had been a change in plans, because the weather had thickened again. The storm had spun its circular tail over Kauaʻi as it passed north and the lingering low-pressure system had brought a new perspective. Thompson spoke to the crew assembled in the dank hotel room. He explained that getting to Nihoa was relatively easy in normal trade winds, but getting back to Kauaʻi could be a problem if winds were not predictably out of the northeast. Even with winds of 10–20 knots, the return required a very long tack to the southeast before the boat could pick up speed to the northwest. The remnant low front from the tropical depression blocked the winds. If and when the low slipped away, would the normal trades return?

we had satellite photos and weather reports to augment the combined senses of five navigators, how did the mariners of old get there? How long did it take and how many lives were lost? Would mine be one of them? I steeled my resolve to endure what might be ahead. I vowed not to puke and not to fall overboard when relieving myself

Thompson polled the crew to determine how much time each one of us could commit to the trip. We each detailed our availability. Frustration and anxiety surfaced as our goal appeared to slip away, but Thompson received the information silently. His demeanor intensified when he spoke again. He had decided to step aside, to not even go on the canoe. Like the low front, his presence blocked the young navigators from making clear decisions. They could only be intimidated by his opinions. After almost 50,000 miles logged on the craft, it was hard for him to let it go, but he left it to the four neophyte navigators to decide where to sail. After the news set in, the crew hunkered down to work, for this training cruise was important for future pan-Pacific expeditions.

Thompson and I left the navigators in the room to plot the course and we headed to the beach. As he stared out at the open sea to sense any change in the wind, Thompson seemed relieved that the torch had been passed. With their leader gone, the apprentice navigators had to make their own decisions and stick to them. Given everyone's schedules, there was not enough time to wait for the sea conditions to stabilize and still reach Nihoa. Besides, the predicted south swell at the island would make landing very difficult. I was secretly glad we were not going to Nihoa, because the responsibility for calling off a dangerous landing would have been mine. The temptation to land on Nihoa after the arduous sail would have been great, but I had noticed the rubber dingy and its small engine on Hōkūleʻa and doubted that it could safely handle the rough surf at Nihoa.

Instead, I dared to believe it was possible that the Hōkūleʻa might visit the Forbidden Island —Niʻihau. From our hotel on Kauaʻi, it was only 35 miles away, yet it seemed a century distant in time. We could sail past the cinder cone of Lehua and then sail around the west side of Niʻihau and circle Kaʻula Rock, a crater rim about 20 miles south of Niʻihau. I once visited Kaʻula by leaping out of a hovering helicopter and spent a day surveying the birds on this 125-acre state wildlife sanctuary. As an albino frigatebird soared over the indigo sea, I imagined that the white wraith might inspire mythic musings, for seabirds played a great

role in the lives of Hawaiians on Niʻihau. They would look across the ocean to Kaʻula to see if the seabirds were swirling and would then know that enemies were staging an attack on their rear flank. In 1932, a lighthouse was built on the summit to warn away ships. Kaʻula than came under the administrative jurisdiction of the U.S. Navy and was used as a bombing range from 1952 until 1976, when endangered Humpback Whales were observed cavorting just offshore. Compliance with the Endangered Species Act temporarily suspended the bombing exercises, but military activities continue to this day.[2]

But instead of sailing to Niʻihau, we sailed to Oʻahu, "the gathering place," to families, jobs, and other demands. Try as we might, we could not change the weather or expand insufficient time. We got stuck between the collective wisdom of the timeless past and the technological dependence of the present. The ancients had lifetimes to voyage, compelling reasons to travel, deep-seated myths to guide them, and an unbroken tradition of seamanship in the vast Pacific Ocean. Even though we also relied on the stars and winds, we had access to global-positioning computers, fathometers, solar panels to power radio uplinks with a satellite lost-and-found system, plus a 50-foot motor sailboat to tow us out of trouble. Still, ancient wisdom

Sooty Tern chick on dud bomb at Kaʻula Rock.

coupled with modern technology could not get us to Nihoa, and *Hōkūle'a* headed for home.

The next day, we said our good-byes to Thompson at the pier, boarded *Hōkūle'a,* and joined hands in prayer. After envisioning a safe passage, the first mate collected everyone's timepiece. This action was ostensibly to prevent anyone from dead reckoning our position, but it was really a ritual used to enter the timeless realm of the ancients. We left Kaua'i under tow from the escort vessel. The sea was running around 8 feet when we cast off our towline and all hands hove to, hoisting the mizzen and mainsail. Wind filled the canvas sails and *Hōkūle'a* took off, cutting through the waves, leaving little wake. The 60-foot twin hulls allowed the craft to span the distance between most waves. Our ride was more stable than that of the sailboat that wallowed in the troughs (a companion vessel always travels with *Hōkūle'a* in case of emergencies). The creaking of the tight lines and the slapping of the loose ones accompanied the rush of water against the hulls. *'Ua'u kani,* or Wedge-tailed Shearwaters, wheeled around us, slicing the sea with their wing tips while Sooty Terns circled higher in the sky. As *Hōkūle'a* sailed through the flock, seabirds plunged into the ocean for fish and squid that had been chased to the surface by predatory tuna.

At sunset, we entered the notoriously rough Kaua'i Channel. Wind and water driven between the imposing hulks of O'ahu and Kaua'i create unpredictable seas—a natural barrier that protected Kaua'i from conquest by O'ahu chiefs of yesteryear. The waves crested and broke at 10 feet, but still presented no problem to the craft. Occasionally, we seemed poised on the brink of a huge maw that would swallow us, but the canoe rose up and floated on the wave crest like a Red Phalarope, a buoyant shorebird that bobs like a cork on the open sea. The fledgling wayfinders huddled in a corner of the craft, reviewing among themselves the site of sunset and the rising positions of stars.

They stared into the twilight and tried to recognize constellations. Starpaths, familiar from their homework, were not yet fixed in the planetarium of their minds. Thompson's teachings found fertile ground in these young navigators. He was present in spirit if not physically, though I wondered which individual would develop the acuity and tenacity Thompson had, for it was legend that he rarely slept during voyages so as not to miss a subtle clue from the environment.

The canoe lurched forward, sliding down the face of a large wave. My stomach followed. Roll and rock, pitch and yaw, constant wind and spray dulled my interest in noninstrument navigation.

Tacking Hōkūle'a.

Wave spume slapped the canvas flap as I crawled into my bunk in the canoe hull to rest, perchance to sleep. My hands ached from handling wet lines, but I had not thrown up—so far so good. Water sloshing against the bulkhead at ear level fell into a tolerable commotion, well shy of the wet pandemonium that would result if the winds came up. Slipping along the edges of waves and consciousness, I felt like an aquanaut drifting to sleep in the first "inner space" craft ever built. All too soon, Tava woke me for my watch. The waning moon splayed light across the waves in a wide, silver swath. Cumulus clouds floated across the moon, and a red shooting star slit the night sky. The ocean slapped the hull while ʻewaʻewa (Sooty Terns) repeatedly cried wide-a-wake-wide-a-wake as if there could be any doubt. The four wayfinders were positioned in the corners of the craft, silently staring into space.

In the dead of night, the universe is the stars, the sea, the canoe, and the mind—the original reality of Polynesians. The jiggy-jaggy day is simplified through the trick of darkness, and primal patterns play out unseen in the black sky and water. I closed my eyelids and my mind's eye glowed as I chanted the mantra Om mani padme hum. In Tibetan it means, "The entire universe is like pure crystal light in the heart of the lotus." It seemed appropriate to chant in honor of the fourteenth Dalai Lama of Tibet, who had recently visited the islands to support Hawaiian Sovereignty. Also, Dalai Lama means "oceanic master" in Tibetan. With nocturnal attention, I opened my eyes and saw a path of moonlight play across the ocean. So much more existed in all other dimensions around me. The moonbeam was like human consciousness in the ocean of awareness. You can see fairly well in the lit areas and even get a sense of the whole, yet the universe is incomprehensibly packed with life and death, opportunities and dangers.

Toward the end of my watch, we spotted a light on the west coast of Oʻahu. Our brief passage across the timeless ocean was over. From now on, we looked for man-made lights and other boats as well as the celestial scenery. The navigators were relieved; they had completed their solo voyage and learned about trusting their senses and intuition. But every lesson leads to another. Unbeknownst to them, as we sailed through the night, an unseen power seized Hōkūleʻa. The strong current near Oʻahu drove the canoe 20 degrees off course. In the harsh morning light, the shoreline was no closer than when we had first spied it.

We were well off our estimated position and needed to tack again. All hands on deck grabbed the lines of the mast. On command, we put our weight to the lines and heaved up the sail, swung the boom to the other side of the deck, lowered the sail, and secured the lines. All this action was coordinated with few commands. Everyone knew the ropes and jumped into a succession of actions to catch the wind. Shortly thereafter, the first mishap occurred. Hōkūleʻa is driven by the sails and steered by a large stern rudder, called the sweep, as well as by two inboard paddles that are manned continuously. The waves that pass under the hull push against the paddles with great force, and a large swell hit the sweep and drove it against the gunwale. An experienced sailor was overpowered by the thrust and his finger was pinched between the sweep and the gunwale. Luckily, we radioed a nearby fishing boat, which came quickly to transport him to a clinic on Oʻahu. The injury was a reminder to be ever alert at sea.

As we neared our destination, the situation grew more distracting by the minute. Soon, an airplane buzzed us at 50 feet, wagging its wings in welcome. Then a tugboat's powerful wake jockeyed the craft as it passed close by to greet us, but it was the fully loaded container ship bearing directly down on us at close quarters in Honolulu Harbor at midnight that caused the most dangerous moments of the entire cruise. On the open sea we were a novelty; in the lanes of commerce we were a nuisance. Did they even see us out of their four-story-tall pilothouse?

As I positioned myself, ready to leave the boat as I had boarded, with a midnight harbor swim, the towering ship swung away. Soon the canoe was tied up, and we gave thanks for our successful crossing by again joining hands in prayer. We then polished off some beers and *'opihi*, miniversions of the saucer-sized ones that occur on Nihoa. The crew dispersed into the night, and I caught a ride to Pālolo Valley where a solid bed awaited. I rued my failed chance to land at Nihoa, but I was glad the trip was over, especially because I didn't have to use the flying-buttress toilet.

References

1. Kyselka (1987).
2. Lipman (1980).

Return of the *Iwi*

E ala e ka lā i ka hikina Awake the sky in the east
I ka moana, ka mana hohonu In the ocean in the deep
Pi'i ka lewa, ka lewa nu'u Rise in the sky to the peak
I ka hikina, aia ka lā e ala e! In the east, Awake!

Blowing of the Triton's Trumpet shell signals departure from Nihoa.

Planted in the collective Hawaiian conscious-ness, the seeds of intention to sail to Nihoa took several more years to germinate. Finally, in November, 1997, another cultural expedition to Nihoa and Necker Islands was organized for the purpose of reinterring ancestral remains. The bone fragments of seven ancient Hawaiians, excavated by Kenneth P. Emory, have re-sided at the Bishop Mu-seum since 1924. During that period, these, possibly the oldest human remains in Hawai'i, have completed their own metamorphosis, from fragments of the secret past, into numbered museum specimens, into part of the foundation of a new sovereign Hawaiian nation.

It had taken an act of Congress (the Native American Graves Protection and Repatriation Act or NAGPRA), passed in 1990, for federally funded institutions to offer to return remains and relics to tribes that can prove cultural affil-iation with them. The Na-tive Hawaiian group Hui Mālama I Na Kūpuna O Hawai'i Nei (Hui Mālama) petitioned the U.S. Fish and Wildlife Service (USFWS), which was the legal owner (Bishop Museum was merely the custodian), to release the remains from Nihoa and Necker. The USFWS was reluctant to surrender the bones without proper and thorough documentation that this particular group was "culturally affiliated" with the remains, a sit-uation played out across the country with other bones of contention. In fact, it took 2 years of legal wrangling before the remains were released to the group. Foot-dragging in the USFWS regional office only heightened the Native Hawaiians' perception that the federal archaeologists were obstructing the final disposition of the bones *(iwi)* of their ancestors *(kūpuna).*

Like *Hōkūle'a,* Hui Mālama is helping to channel the energy of the ancestors into a new Hawaiian nation. The group is rebuilding the foundation stone by stone, bone by bone. Edward Halealoha Ayau, the legal counsel and trip organizer, voiced the belief that the individual's strength is not in the flesh and blood, but in the bones. The bones hold the *mana,* the spiritual energy, in place in the *'āina,* or land. When the ancestral remains are removed from the *'āina,* the connection to the living present is broken. The removal of remains and casual handling of bones by "non-Hawaiians" during the postcontact period sapped the vitality from the living Hawaiian culture and it slowly declined. The key to regaining power and strength is to make the land whole again; in repatriating the archived ancestors, the foundation can be rebuilt and made strong so that the revitalized native culture of Hawai'i can stand tall. It is on this reinforced foundation that the new sovereign nation will be built.

It all fell into place in November 1997. Hui Mālama chartered the *Stardust,* a 50-foot sailing yacht, to carry the *iwi* back home. They also had to obtain a special use permit to visit the remote islands in the National Wildlife Refuge. The permit stipulated that they take an

Mo'o Head, Necker Island. Mo'o *are dragonlike demigods able to change their body forms at will.*

escort, and I joined the ship as a pilot again because I knew the waters around Nihoa and Necker Islands. As a representative of the U.S. government,

I felt like the "occupying force," the previous possessor of the ancestral remains. I joined the seven Native Hawaiian crew as an outsider, slightly depressed at the prospect of sailing into the November high seas in cramped quarters. Yet, I was thrilled and considered myself privileged to make this historic trip. I felt chosen to complete the cycle begun by Emory and the Tanager Expedition, who had unearthed these human remains almost 75 years before.

Indeed, there was a sense that fate had chosen all of us to be together at that time to perform that deed. The idea that we were guided by spirits of the past came up time and again, especially when nonpartici-

Sunset behind Northwest Cape, Necker Island.

pants queried us on the weather situation. "Isn't mid-November kind of late in the year to be sailing into the open seas?" The answer was always the same. The process to return the *iwi* had begun years before and led directly to this point in time. The spirits in the bones had been summoned from their slumber and were helping to direct events. When it was time to land on Nihoa and Necker, they would be doing their part to get home.

Faith was the order of the day, and I had an immediate demonstration of this power before we set sail. My friend Mukunda and I went out to lunch and when we returned to the pier, the sailboat had been moved to another dock. The crew had waited for my return to tell me where the new pier was located. I jumped out of Mukunda's car and quickly said good-bye. He drove off and I got into my van and followed the crew's vehicle through dense rush-hour traffic. As we crept down Ala Moana Boulevard, an unsettled feeling gripped me. Then I realized that my backpack

filled with cameras and my books were still in my friend's car!

Locked in rush-hour traffic, I realized that I had to make some radical traffic moves to exit immediately and go to Mukunda's house to wait for his return with my pack. I signaled my intentions to the *Stardust* crew and then moved glacially to the far left lane to exit. In a mental muddle, I replayed the possible scenarios about how this was going to turn out. When I looked at the crew's car to gauge my progress, I saw that Mukunda's car was directly behind it, right where I had just been! He was waving at them to stop. He tried to give them my pack, but they pointed to me two lanes over and shook their heads in disbelief. Shocked by the synchronicity, Mukunda ran to my car and silently handed me the pack. Somehow he had realized at the same time I did that he had my pack. But how he had managed to materialize behind us was the miracle that christened the trip for me.

A long day of anxiety was about to end. The leaders of Hui Mālama finally arrived at the

Stardust with the *iwi* resting in woven baskets wrapped in black cloth. We quickly gathered together for a ceremony of departure. Holding hands in a circle, the entire crew and support team chanted and prayed for propitious weather and a safe, successful voyage. Lines were cast off an hour before sunset and we set sail for Nihoa. Under way finally, a supper of fried chicken was served. We were hungry after all the stevedoring we had done, and it was heartily consumed. But before the night was over, there was silent suffering among the crew without sea legs. We cursed the fried chicken through our teeth. The sturdiest sailors manned the boat through the Kaua'i Channel that night while the trade winds blew at 15 knots and the swells rolled in 6-foot sets.

The next day we breezed past the Garden Isle of Kaua'i wreathed in clouds. To the south, Ni'ihau sat low in the water. As we approached, we saw the islet of Lehua, an eroded cinder cone 702 feet high about a half mile north of Ni'ihau, home to seabirds as well as introduced rabbits. Swells crashed on this crescent cone and I worried about our future landing at Nihoa, now a day's sail away. We were passing through Kaulakahi Channel, from the high islands into the watery realm of the islands under the horizon. The back gate between Lehua and Ni'ihau closed with the arrival of nightfall, and a stiff wind drove our funerary barge dead downwind.

The next day was D day at Nihoa. Representing the Feds, I had the final word on landing and felt responsible for the ancestors' homecoming. It was pretty rough out there and a little less wind for the landing would be nice. I tossed and turned in my bunk. The rush of water next to my head and the pitching of the craft terrorized me about how "out there" I was. Joseph Conrad called it The Shadow-Line, the realm of stars, sun, sea, light, darkness, space, and great waters. It is the formidable Works of the Seven Days, into which mankind seems to have blundered unbidden. I

managed to doze off but was up for the 3 to 5 A.M. watch. I stood watch with two others in silence as a cold wind drove us west.

At first light, Nihoa rode the horizon barely visible in a sea of blackness. Her jagged summit snared the tentative rays. As the sun illuminated the rock with orange light, flocks of black frigatebirds, or '*iwa*, flew out to greet us. This pleased the Hui members because the birds are symbolic ancestors. Eddie Ayau even had a bird flock tattooed on his torso. The '*iwa* circled the boat, beckoning back the *iwi*. As we approached the rocky island, the wind was shifting to the east. Northern swells hit the buttress of Nihoa and wrapped into south-facing Adams Bay. But although surf crashed on the lava benches, it did not preclude a landing. It looked dangerous but doable. Like sirens on the rocks, the fan palms waved us in with their gigantic hands. It reminded me of the campaigning politicos who stood on street corners during every Honolulu election and waved foam-rubber *shaka* signs to passing motorists.

When quizzed by the crew as to when I was last here, I had to admit that it had been over 15 years before. That's when Uncle Les took over. He and Lopaka were adroit boatmen, used to landing on the lava apron of Kaho'olawe, an island off Maui. After I pointed out the traditional landing spot, I was dismissed—no longer a participant in the decision-making process, although I was still a member of the landing party. Although it was a bit of a shock to my ego to be so abruptly demoted, Les was correct in doing so. Time was of the essence and the need was great to get the remains ashore and reinterred as deftly as possible. I lent a hand when needed but stayed out of harm's way. The captain maneuvered the *Stardust* under the lee of the west cliff, hoping that some protection could be found. In the rocking swells, a rubber Zodiac was inflated and lowered away. The 30-horsepower engine was lowered into the boat using the ship's halyard. This was a critical point, because the heavy outboard was swinging dangerously from the boom.

Uncle Les and Lopaka wrestled the engine into the boat and onto the stern. Another, smaller backup boat was also inflated, so that in case of an accident it would be ready to go. Then Les and Lopaka set off to survey the shoreline and seek the safest landing place. Les found a spot where the swell was diminished and set up an anchor-and-buoy system. The anchor held a polypropylene line that was also tied to the shore and could be used to pull the boat in. The landing party could exit and hold onto the line, and the boat could reverse if a swell threatened to dash them on the rocks. This clever system permitted us to get ashore safely in rough water.

I handed out pristine "tabis," two-toed, felt-bottomed reef walkers, to assure that no seeds were inadvertently carried ashore in our old shoes. We had taken precautions in Honolulu to freeze all of our clothing as well as my camera gear to kill any living stowaways. However, I noticed a polished staff of wood on the deck that had not been frozen with the other gear. When I approached the spiritual leader of the Hui to inquire about the role of the staff in the ceremony, he said he intended to bury it with the remains and seemed to resent my intrusion and presumptions. Yes, he was probably right that nothing was living in the wood, but he was possibly wrong, and a wood-boring beetle that emerged from the staff might infect the endangered Nihoa palms. After all, about twenty new species of insects are introduced to Hawai'i each year, many having disastrous effects on local agriculture and native flora. That is more than a million times the natural rate [of introductions] and more than twice the number absorbed each year on the Mainland, according to U.S. Senator Daniel Akaka of Hawai'i. I made my plea on behalf of the Nihoa palms, and we reached a compromise that the ceremonial staff could go ashore but had to return to the boat.

Uncle Les had rigged the landing perfectly. I grabbed the lifeline and pulled the boat toward shore while the engine idled in neutral. Timing my jump to the wave slack between surges, I leaped onto the rocks as the boat reversed. The landing party assembled one by one on the rocks, gleeful to be here, where visualizations met reality. But there was no time to linger because the sea state could change at any time. I was asked to leave the funeral ceremony after we had discussed where to regroup. As I set off inland, a Nihoa Finch was there to greet me, drinking from a tiny water seep where a cushion of moss about the size of a quarter grew in the damp nook: a major wetland habitat on this arid island.

I headed up Middle Valley to a cliff where the closest ruins were visible, thinking the Hawaiians had agreed to follow, but they ditched me and went to another area. In a cynical moment, I

Noio *(Black Noddy) and glass ball on sea surface, Beaufort 1 conditions.*

thought that the ancestors must be really angry with the Hui. The final wishes of the deceased were probably to rest by a cool waterfall surrounded by kukui trees on Kaua'i. I consoled myself by peering over the cliffs at the head of the valley on the north face. This is Mauloku, the leaping place of souls, and I briefly thought of merging with the infinite, exchanging places with the souls now returning to their place of rest, but then I thought again—better get back.[1] The tide was rising and the surf was up. I saw the Hawaiians returning from their ceremony, so I also headed to the landing cove. After a final blast from a *pū*, the shell of a Triton's Trumpet, we were ready to vacate the premises. Getting off was wetter but well done and accomplished before noon. All of us were greatly relieved that the ancient Nihoans were finally laid to rest. I called the USFWS on the satellite phone to pass the news. After searching eighteen or so satellites to make the best connection, our message reached Honolulu and news of our success was passed to family members of Hui Mālama and concerned federal officials who had monitored our progress.

Magenta gray and orange clouds turned pink gold, and Nihoa looked like a dark door in the horizon. Only 12 hours separated us now from Necker, and I thought how the ancient mariners, having traveled this far, might easily have stayed the course to reach Necker Island and then pushed on 75 miles to French Frigate Shoals. It seemed that the entire northwestern half of the Hawaiian Archipelago was within their grasp. But one day at a time. So that I could be on watch at 3 A.M., I showered and turned in. Working the graveyard shift, I witnessed black mushroom clouds spilling quicksilver in the full moonlight. During the watch, we crossed the tropic of Cancer at 23° N latitude and left the Torrid Zone in our wake.

By midmorning, Necker Island appeared in the distance. On closer inspection, I recognized the *mo'o* crouched like the Sphinx of the Sea. I looked for its spines and saw rows of *marae* along its back.

Its haunches were streaked with layers of red lava. Its face looked like a Marine Iguana of the Galápagos. Whitewash on bird perches became gleaming teeth, and a lava plug, its beady eye. The chin was underslung and venomous drool churned in the surf. Fantastically, the geologic feature has no geographic name: thus I propose here and now that it be called Mo'o Head!

We could see the shrines along the crest of Necker. To the crew, they symbolized pushing the limits and proving one's sailing ability. Perhaps the rigid structure of the ancient class society in the main islands may have forced independent-spirited people to set off on their own and seek new lands. Each voyage might have required a shrine and could account for the thirty-three *marae* on Necker. I especially noticed one visible through a cleft in rocks of the Northwest Cape where it meets the main island. It was conspicuous on the opposite hill and seemed to shout "No fear!"

By noon, as we prepared to go ashore, a gusty north wind was blowing. The high tide and swells made landing at the traditional site in West Cove impossible. Humping swells broke on the lava ledges, sending plumes of spray up the cliffs. I thought it was too dangerous to land safely and said so. I certainly was not going ashore. "That's right," said Uncle Les, "only one or two will go—if possible." The Zodiac was carefully lowered away into the confusing sea and Uncle Les and Lopaka motored off to inspect the ledges and time the wave sets. They returned with the news that a 15-second interval of slack water existed between surges and that it was possible to dash ashore in that period.

Quickly, the selected landing team made ready to go. The *iwi* were carefully handed to them in the waiting boat. Maka, an expert waterman and *kanaka maoli* (Native Hawaiian), had no problem getting onshore. Puna, the priest or *kahuna pule,* also made it ashore easily with his important package. They climbed up the cliff face along a basalt dike that offered a staircase to the island summit.

Along the narrow ledge an albatross stood in their way. But the *mōlī*, symbol of Kāne, presented no threat when the two men skirted it. We anxiously watched their progress through binoculars. As the reinterrment of the Necker remains proceeded, the crewmembers in the pickup Zodiac decided to circumnavigate Necker. At one point, the Zodiac disappeared from our view on the mother ship at the same time that the emergency boat we had in the water bumped the *Stardust.* The impact knocked the small outboard engine into the drink. For several nervous minutes we waited for the Zodiac to round the other end of Necker and assure us that the entire party could make it back safely.

They did it in grand style. All the ancestors were home, the historic mission was successful, and no one got hurt. The Hui's efforts, spanning 2 years to obtain and reinter the ancestors, were finally completed. Uncle Les called the jubilant crew together to discuss the event. Everyone had a say about what they felt and thought about this historic day. A feast accompanied the straw yellow moon rising in the pink and blue sky. After a prayer to Kanaloa, god of the sea, we pulled away from the mesmerizing spot. Kanaloa was evidently pleased, and the sea surface said so. We set sail into a realm of diminishing winds and swells. Having paid our dues, a feeling of peace overwhelmed us because the dreaded return passage into the driving winter trade winds had been waived.

The cobalt blue seas were glassy, with only the slightest ripple. Billowy cumulus clouds glinted on the cyanic mirror. A pollenlike scum, said to be from tuna spawning at the full moon, marred the burnished surface. I could even spot the Pacific Pelagic Water Strider, *Halobates,* the only oceanic insect, skipping away from the wake. With no wind, we had to motor at 7 knots, heading toward Weather Buoy 51001 to do some fishing. This buoy detects and transmits the arrival of north swells and alerts the main islands that surf is on the way to the northern shores. It also is an FAD (Fish Aggregating Device), an oasis where a food chain builds up on the 2-mile-long anchor chain. This buoy was booby heaven, judging from the accumulations of guano. All three booby species utilized the perches, and flocks of Black and Blue-gray Noddies swarmed over the shoals of tuna driving frenzied prey to the surface. We caught a couple of *aku* (Skipjack Tuna) and departed for home. Raw and grilled *aku,* rice, and salad made a sumptuous supper. When I came on watch, the moon was burning bright, but when the light was muted by a cloud, I saw the sea foam glowing electric blue. Microscopic one-celled animals, called bioluminescent dinoflagellates, flared in the boat wash. The effect resembled the blue-light bug-zappers situated around outdoor restaurants in town.

Buoy 51001, pilfered of high-tech gear, is a booby perch.

Masked Boobies in silhouette.

Coincidentally, after we had passed through the area, the U.S. Coast Guard buoy-tender *Sassafras* arrived that night and apprehended a fishing boat tied up to the buoy. The Honolulu papers later reported that the boat was searched for solar panels and other gear stolen from the buoy. No gear was located, but the ship was fined for tying up to the buoy. The Coast Guard replaced the buoy during its stay, but within a week, the $160,000 buoy was robbed again for its high-tech parts. Within 3 months, the buoy was plundered three times, and now the FBI is investigating.[2] The pirating of federal property that informs the boating, surfing, and fishing communities of north swells and impending foul weather is a bad sign of the times. The USFWS has taken heed of these acts. In the face of potential wanton destruction, it plans to mount a security camera on Nihoa to record illegal trespass to preserve the biological integrity of the island.

We sailed into Honolulu on Sunday morning, 6 days after we had begun our expedition. The final prayer at the dock was especially poignant. We recognized that the power of the past could become manifest in the present and influence the future. The ancestors did their part to summon forth the bearers of the bones, to push us along, and calm the seas for our safe return. Faith in the *mana* of the *iwi* was enough to fuel a new sense of sovereignty and reverse racism in me. I was en-vious that the Hui members could claim connection to the first Ha-waiians. My own ancestors, who came from Yugoslavia in the midst of civil war, were so alien to my present existence that I felt more kindred to the Hawaiians. But that was cultural ap-propriation. Indeed, my thinking, writing, and speculating was verging on cultural thievery, some-thing we *haole*s are renowned for. Yet my past re-quired me to plumb its own power. I had met old Emory, shaken his spidery hand, read his works, and now had accompanied fruits of his endeavors back to their resting places. I was connected to him as I represented his contribution to science (now deemed politically incorrect in some circles). But the past is rectified and sanctified by the *iwi* of the *kūpuna* back in the *'āina*. I learned that the life of the land is indeed perpetuated in righteousness.

References

1. Pukui et al. (1974).
2. TenBruggencate (1997).

7

The Coral Kingdom of La Pérouse

He pūkoʻa kani ʻāina.

"A coral reef that grows into an island."
A person beginning in a small way gains steadily until he becomes firmly established.

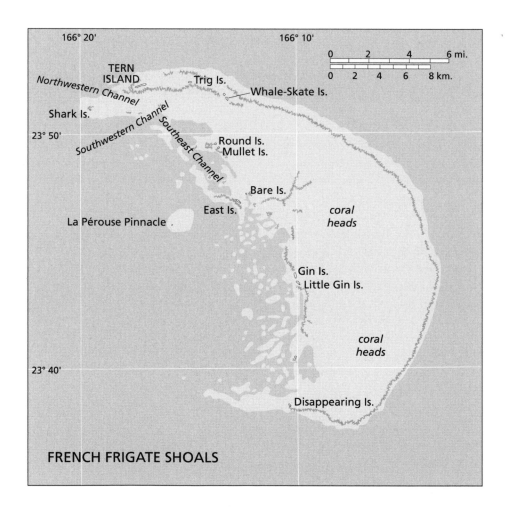

Map of French Frigate Shoals.

French Frigate Shoals is the midpoint in the Hawaiian Archipelago. It is also the first atoll in the Leewards and the largest coral reef area in Hawai'i. Its almost 200 square miles of coral reefs, combined with Maro Reef and Pearl and Hermes Reef, totals 500 square miles, while the combined reef area in all the main islands is only 360 square miles.[1] The reefs are primarily composed of the coral *Porites,* cemented together with calcareous algae. These northernmost coral reefs in the world are relatively depauperate compared with southwestern Pacific reefs, which lie at the center of aquatic biodiversity. For example, the *Acropora,* or "table-top" coral, community of French Frigate Shoals is unique in Hawai'i, yet *Acropora* corals are the largest group of reef-building corals worldwide. There are at least one hundred species of this genus, representing about one-fifth of all corals, yet only two occur in Hawai'i at French Frigate Shoals. In addition, there are over twenty other coral species in the fabulous underwater gardens at French Frigate Shoals.

About 700 species of fish, 400 algae, 1,000 mollusks, and 1,350 of other invertebrate groups inhabit the inshore waters of Hawai'i.[2] The fish and macroinvertebrate faunas of the Leewards are generally the same in species composition as those of the main islands. Most of the species are also distributed throughout the Indo-Pacific region, but approximately 20% are endemic to Hawai'i.[3] For example, one species of butterfly fish is endemic to the unique *Acropora* coral community of French Frigate Shoals. Hawaiian monk seals infrequently appear outside the northern reef ecosystems.

A coral reef is an oasis of life in a relatively barren ocean because it provides a wide variety of habitats: submerged banks, fringing reefs, sand flats, rocky shoals, and coral sand beaches. The bird's-eye view of French Frigate Shoals is breathtaking. In the aquamarine lagoon, yellowish heads of cauliflower-shaped coral shimmer under transparent turquoise water. Rust-colored fringing reefs, streaked with foam, protect the crescentic sweeps of sand on the north and east sides. The green vegetation of East, Tern, and Trig Islands, which compose the land portion of French Frigate Shoals, soothes the eye, and the blinding white islets of Disappearing, Gin, Little Gin, Bare, Round, Shark, and Mullet Islands gleam on the palette of blues. La Pérouse Pinnacle stands black and white before the deep blue sea.

These islets of French Frigate Shoals may be the inspiration for the name Moku Pāpapa, found in Hawaiian legends. In 1778, Captain James Cook recorded the following comments in his journal: "...we got some information of a low uninhabited island in the neighbourhood of these [main islands] called Tammata pappa."[4] After Cook's death in February 1779, the journal of his replace-

Aerial photo of Whale-Skate Island, French Frigate Shoals, 1990.

ment, Captain King, also mentioned this island: "To the WSW of Teula [Kaʻula], they visit a low sandy island for Sea birds & Turtle called Modoo-papapa or Komodoo papapa." Another footnote stated: "One canoe belonging to some Atoui [Kauaʻi] Chief staid with us till Sunset, and then went towards the Island Outoura [Lehua] which was 4 miles distant to the SE. Their business, they told us, was to catch red birds, and the next day they intended going to Tomogoopapappa for Turtle." King also related that his efforts to locate the island were unsuccessful. Beaglehole (1967) wrote in a footnote to King's comment: "This 'low sandy Island' is quite baffling, though the name was picked up both in 1778 and 1779. [Ka] motu or moku papa is literally the low, or flat and smooth island.... Dr. Emory suggests that, as Nihoa was known to the historic Hawaiians and frequently visited by them, Moku Pāpapa may have been an alternative name for this island."[5]

The literal meaning of *moku pāpapa* is an island that is low and sandy. This suggests that Emory was wrong. Nihoa is certainly not flat and sandy. The nearest island that fits the description, and is west southwest of Kaʻula, is Johnston Atoll, about 800 miles away. However, French Frigate Shoals lies only 400 miles away. Aside from being in the opposite direction from what the sailors reported, it better fits the bill because it is where over 90 percent of the sea turtles in Hawaiʻi nest. This must be Moku Pāpapa, where sea turtles can be easily picked off the sandy beach, not hefted into canoes from the rocky shelves of Nihoa or Necker, where they are relatively uncommon. Also, *moku pāpapa* is the generic atoll name remembered in ancient chants and legends. It may represent way stations on the voyages to Kahiki. In the reign of King Kalākaua in 1886, it was the name used for Kure Atoll, the last Hawaiian island. We will never know the extent of wanderings of the prehistoric Hawaiians, yet it seems reasonable that French

Frigate Shoals, only 75 miles from Necker, would have been within their reach.

Basse des Fregates Françaises, "Shoal of the French Frigates," was (re)discovered on 6 November 1786, "almost by accident." Two French frigates, the *Astrolabe* and the *Boussole* under the command of Count La Pérouse, narrowly averted running into the reef. The crews' ability to come about in less than a quarter mile saved the ships. "The moon, which was almost at the full, gave so great a light that I thought we might venture to stand on.... Since our departure from Monterey, we had never experienced a finer night, or a more pleasant sea: but the tranquillity of the water was among the circumstances which had nearly proved fatal to us. Toward half an hour past 1 o'clock in the morning we perceived breakers.... From the smoothness of the sea, they hardly made any noise, and some foam only, at distant intervals, was perceptible. The *Astrolabe* was a little farther off, but she saw them at the same instant with myself. Both vessels hauled on the larboard, and stood with their heads south-southeast; and as they made way during their maneuver, our nearest distance from the breakers could not, I conceive, be more than a cable's length."[6]

A natural monument in the atoll honors the event. La Pérouse Pinnacle, a 1-acre plug of exceptionally dense lava from the throat of the original volcanic cone, is all that is left after the surrounding rock has eroded away. It is the atoll's navel, an "earth mark" of its fiery birth. This whitewashed rock stands alone in the southwest quadrant of French Frigate Shoals, 122 feet above the sea. From a distance, it looks like the *Astrolabe* under full sail: the canvas luffing in the freshening trades, with the treacherous reef dead ahead.

Indeed, French Frigate Shoals has seen more than its share of wrecks by others less keen than the Count. The whaler *South Seaman* was lost at French Frigate Shoals on 13 March 1859 while en

Seen from a distance, La Pérouse Pinnacle resembles a sailing ship.

route to O'ahu. With a few gallons of water and some hardtack, the crew set off in a whaleboat for Guam, but luckily they were saved by the Hawaiian government ship *Kamehameha IV* sailing in the region.[7] On 14 April 1867, the weather was beautiful with a full moon and the whaling bark *Daniel Wood* was steadily pursuing her course with a fine favorable breeze until "Hard up the helm!" split the night. *Daniel Wood* wrecked and the captain and some crew set off for O'ahu for help. After an open-boat voyage of 8 days on a pint of water and a biscuit apiece a day, they reached Honolulu and sent aid to save the stranded crew.[8] To survive, the thirty marooned men had killed over one hundred turtles before being rescued.[9]

In 1872, the vessel *Kamehameha V* found some odd survivors of that shipwreck. On 4 July at French Frigate Shoals, they recorded "two large hogs on a sand spit, a quarter mile in circumference. They have been there since April 1867. There is no fresh water and very little vegetation. As soon as the boat landed, the hogs took to the water and swam off to some rocks just awash, and seemed perfectly at home in the water."[10]

One hundred years later, ship cargoes proved to be a greater threat to the reef. In 1977, billows of black smoke a mile high towered over the 846-foot, Liberian-registered oil tanker *Hawaiian Patriot* burning 200 miles north of the atoll. The tanker exploded and sank, releasing an estimated

Sooty Terns, the most abundant bird in the central Pacific Ocean, do two things almost continuously: cry and fly. Only at the breeding colony do they descend to earth, and even then they are rarely quiet. (Drawing by John Gilardi)

five million gallons of light crude oil in what was, at that time, the worst oil spill in history. Luckily, the swath of oil, 14 by 40 miles in extent, largely evaporated. Little oil washed ashore, but some birds were oiled at sea. In 1980, the Greek freighter *Anangel Liberty,* carrying 2,200 tons of kaolin, a clay used in making porcelain, ran aground at the Shoals. To get her off the reef, the fine powder was dumped overboard. Again, the French Frigate Shoals coral reef ecosystem, the richest in Hawai'i, escaped with minor damage. In 1982, the fishing vessel *Keola* hit the reef and sank at Gin Island (named by Dr. Wetmore of the Tanager Expedition for the fond memories recalled by an empty Gordon's Gin bottle found on the beach), stranding the crew of four for 11 days and spilling 8,000

gallons of fuel. The boat was on autopilot with no one at the helm when it struck.

My arrival at French Frigate Shoals in 1977 was less dramatic. A blizzard of birds filled the air as the C-130 cargo plane touched down on the runway at the Tern Island Coast Guard Station. As my eyes adjusted to the blinding midday light reflected off the white coral runway, I saw thousands of screaming terns whisking away from the thunderous plane. Landing on this short, rough runway is a pilot's nightmare. A tern sucked through a turbine engine could down a plane, and the consequences of an albatross going through the windshield were unthinkable. The wing of our plane had hit a bird and the pilots were mad as hell because a dent in the wing can cost thousands of dollars to repair. Besides, the pilots might get stuck at the worst duty station, by some accounts, in the entire Coast Guard system.

A year on the 56-acre rock, with millions of noisy birds and twenty other stir-crazy guys, made the men of Tern Island desperate for their bimonthly mail plane. Angry pilots were to be appeased at all costs. The sunset red yolks from the terns' smashed eggs oozed into the coral gravel every 2 weeks. The Coast Guard received permission from the U.S. Fish and Wildlife Service (USFWS), which managed the island as a wildlife refuge, to haze the terns from the tarmac edge where they nested. Still the pilots were not satisfied —there were too many other flying objects. During peak breeding season, the planes avoided landing altogether and air-dropped canisters of mail, which either landed in the sea or shattered poorly packaged materials upon impact. During my 3 months on the rock, I was the public relations man for the birds—a tern interpreter. I tried to explain that Tern Island is for the birds, but Hitchcock's movie *The Birds* haunted everyone's imagination.

As the flock gathered together from across the tropical Pacific Ocean, the swarm of Sooty Terns swirling over Tern Island grew denser each night.

Sooty Terns nesting on the runway apron at Tern Island. The terns nest simultaneously in large colonies, providing more food than predators can eat, ensuring that some chicks will fledge.

Constantly calling *rad-a-rat-widdeyap-rad-a-rat,* the tornado of terns was preparing to breed and their flocking behavior helped to synchronize the colony. Suddenly, two terns rocketed out from the flock. In tight formation, they glided swiftly, separated by only a wing length. They soared together, forging a pair-bond in flight that will last the breeding season. Each day, more and more pairs jetted out in formation, and each day the cacophony grew louder as the flock of terns slowly descended closer to earth. After several weeks of aerial coordination, the terns landed on the runway, more or less en masse, and laid their eggs within days of each other. This limited the entire population's time on land, shortening the amount of time they might be susceptible to predation.

Sooty Terns are the most abundant species of seabird breeding in the central Pacific. They are also the most numerous birds nesting in the Northwestern Hawaiian Islands. Almost one-and-a-half million breed here, with at least as many non-breeders present. Sooty Terns have evolved an aerial existence. They rarely risk an ocean landing because their plumage is not waterproof and their feet are too weak to propel them off the surface. Demonstrating their mastery of flight, they can sleep on the wing. Spending up to 9 months of the year in the air, they fly high-altitude circles on

Sooty Terns breeding at Tern Island.

"autopilot" while napping. No one knows how they do it, but they may act like some gulls that show sleep brain waves in one hemisphere of the brain at a time. They may also have "minicomputers" in their heads. Scientists have isolated bits of magnetite imbedded in their brain cells. These "chips" may help the terns orient to the geomagnetic force field in the upper atmosphere and thus navigate across the trackless Pacific.

Ironically, or perhaps even typically, it was humans who made Tern Island so attractive to Sooty Terns. Today, it hosts almost 80,000 breeding pairs of Sooty Terns, making it the third largest tern colony in Hawai'i. However, only small numbers of terns nested on a 6-acre sand bar called Little Tern Island in 1940. In August 1942, Seabees, the men of Company B, Fifth Naval Construction Battalion, extended the islet from 1,880 feet to 3,100 feet long and standardized the width at 350 feet with steel sheet pilings holding 660,000 cubic yards of coral fill dredged from the surrounding reef.[11] Ground-up coral mixed with sand and phosphate topped the runway, and under constantly wet conditions the runway chemically fused together into coralline

rock. In 9 months, the job was complete at a cost of nearly two million dollars. Today, the seawall is rusting away and estimates to repair it run over twelve million dollars.

Tern Island, which resembles an aircraft carrier under way with its white wake flagging behind, is eroding away. The cement pad where the diesel tanks once stood is gone. The southwestern beaches are disappearing. In 1997, the USFWS had to place many tons of imported lava rock to save the former Coast Guard barracks. Preserving the footings of the structure is equivalent to maintaining a presence on the island. If rising sea levels are the wave of the future, Tern Island may yet return to the 6-acre sandbar it once was, regardless of the dollars spent.

The rush to build Tern Island was in response to an act of war. The Japanese had secretly used the lagoon at French Frigate Shoals for trial runs on Pearl Harbor.[12] Two Japanese seaplanes landed and refueled from submarines on 3 March 1942. After refueling, the planes flew a nighttime reconnaissance of Pearl Harbor and dropped some bombs in Punchbowl Crater before hightailing it back to a Marshall Islands base.[13] Later, three Japanese submarines were sent to French Frigate Shoals to refuel planes staging a second Pearl Harbor reconnaissance. But two U.S. Navy seaplane tenders were in the lagoon when the Japanese visited in late May, thereby denying the Imperial Navy knowledge of the location of the U.S. naval fleet at a critical juncture. In fact, the fleet was then north of French Frigate Shoals, heading to Midway for what proved to be the pivotal battle in World War II.

Up to twenty-two U.S. seaplanes used French Frigate Shoals as an anchorage and flew 100-mile-radius reconnaissance flights daily. Only seven

U.S. Navy biwinged floatplanes and ships at East Island, French Frigate Shoals, USS Oglala *and La Pérouse Pinnacle in background, 28 April 1933. This coordinated ship-plane exercise, consisting of thirty seaplanes and seven ships, began at Pearl Harbor and then moved on to Johnston Atoll. (U.S. Navy, National Archives, Pacific Sierra Region)*

Seaside Fishing Company and the Hawaiian-American Fisheries Company used the runway and docks to fly fish and sea turtles to market.[14] Yet, the loran station stayed on East Island despite tidal waves, delinquent mail, and canned food until 1952, when the operation moved to Tern Island. One antenna pole was left standing as a landmark that you can just make out as a vertical pole from most parts of the lagoon. It marks a refuge from the sea, as noted by a plaque on the pole, which reads:

enemy ships were seen during the remainder of the war because the action was elsewhere in the Pacific theater. As the Navy planes were cooling down on Tern Island, the air waves were heating up on nearby East Island. By the end of July 1944, a crew of twenty-seven Coast Guard personnel moved into their new Quonset huts and went on the air with a Long-Range Navigation (loran) transmitting station. Low-frequency radio waves transmitted from seven pole antennas on East Island pulsed out signals that, when triangulated with other loran signals, enabled mariners to locate their position on the ocean.

By the end of the war, the salt and wind began to corrode the electrical equipment and erode the men's morale. Flights to Tern Island had been reduced to one per week, and East Island requisitions went unfilled. In 1946 the Navy pulled out of Tern Island and commercial fishermen of the

Walk softly, stranger.
The land on which you stand
Is Holy Ground.
For here, where seabirds make
  their home—
Men of the Coast Guard once called it home.
From here, a signal pulsed to guide the lost
And weary traveler far from home.
And though this silence—broken now by sounds
  of birds—
Gives no hint of voiceful mirth and laughter;
Yet, to those long gone, it was home—away from
  home—
A place of unspoiled beauty, colored
By the hand of GOD.
And you who stand upon this land
Will someday too,
Remember sunwashed sands and quiet days
And moments crystallized in time.
Walk softly stranger—
For you stand on Holy Ground.

U.S. Navy men building a temporary small boat dock at East Island, French Frigate Shoals, 22 April 1933. An extensive "tent city" was later constructed for naval exercise purposes. (U.S. Navy, National Archives, Pacific Sierra Region)

tifying the structures to withstand heavy seas, the walls were built to "go with the flow." And just in time, because in early December 1969 waves reported to be 50 feet high broke over the fringing reef. Rushing water knocked out the walls without damaging the superstructure and only the electrical equipment was swamped. As the Guardsmen clung to the top of their dormitory to avoid being shark chum, helicopters airlifted them to safety. Two months after the storm, however, French Frigate Shoals was back on the air again, providing locational services to an increasing number of fishermen as well as lodging and support to scientists such as those from the Smithsonian Institution's Pacific Ocean Biological Survey Program (POBSP).

The POBSP was a cadre of biologists reconnoitering the remotest atolls in the vast Pacific, gathering data on plants and animals and placing metal bands on millions of seabirds. Their obser-

Dedicated to the brave men who manned the Coast Guard LORAN station on this, "East Island" spot—called: Gooney Bird Island—from the year 1944 through 1952.

As I read those lines and gazed over the translucent lagoon, I was glad someone had waxed lyrically, honoring inspirations felt by the many men who had passed this way.[15] But if they could only see the place now! Indeed, their "holey ground" is riddled with shearwater burrows and pockmarked with sea turtle nest pits.

In 1952, the Coast Guard moved to the Tern Island loran facility and set up shop. By 1964, they refurbished the buildings with hurricanes in mind. Instead of for-

U.S. Coast Guardsmen smashing tern eggs on the Tern Island runway apron, April 1977.

vations, collected during ten trips to French Frigate Shoals in 1963–1969, help wildlife managers today appreciate the importance of long-term monitoring in island ecosystems. Several scientists of the POBSP drew together all that was known of the Northwestern Hawaiian Islands and published histories and species accounts for each island in the northern chain. I have drawn from their efforts recorded in various issues of the *Atoll Research Bulletin* with thanks and profound admiration for their thorough work.

I was among the second wave of researchers hosted by the military and noted that the life of the enlisted man had changed little in 25 years. Videos had replaced two-reel movies, pinup girls had come and gone, but the isolation of French Frigate Shoals had remained the same. Periodic phone calls had to be patched through ham radio or maritime operators with the entire Pacific listening: "Hi, Honey, it's me . . . over. . . . When are you coming home? . . . over. . . . What? . . . over." Moreover, with Honolulu 480 miles away and only 2 weeks vacation a year, most men slowly went crazy. Our entertainment consisted of drinking warm beer at the "Playboy Club" on the beach, basking in the radiation from the sun and the loran tower (a fluorescent light bulb pointed at the tower would flicker from the incipient radiation like the northern lights).

Time went by relatively quickly, with each of us having his own room and role in the daily grind. Every day, the engineers maintained the generator so that the loran stayed on the air and made sure that the air conditioning, refrigeration, and the salt water desalinator functioned properly. The electronic technicians kept tuning the vibrations, and the cooks prepared the vittles. A junior lieutenant fresh from the Coast Guard Academy ran the show. This diesel-based lifestyle was maintained by about twenty men, whose average age was about 20.

The callowness of the Coast Guard crew did present some problems. On April Fools' Day, I had a not-uncommon adventure for French Frigate Shoals. I was relaxing on my day off at Tern Island when a Mayday call was received. Some young Coast Guardsmen had taken the Boston Whaler out fishing near La Pérouse and were having engine trouble. The only functioning boat was the USFWS's Boston Whaler, a boat whose weaknesses I knew only too well. Some misplaced sense of responsibility took hold of me (I was 25 at the time), and I volunteered to pilot the craft to rescue the Coast Guard. Four of us rapidly deployed the Whaler and sped out to La Pérouse Pinnacle, the last point of contact for the group.

When we reached the outer shoals, the seas were no longer constrained by the atoll platform and were free to heap higher. We were in the open ocean now with squalls passing over. Heading in

Bones of the island, Tern Island seawall rusts in peace.

Le visage de La Pérouse.

La Pérouse Pinnacle stood at the edge of the shoals and watched our progress, which appeared nil. We could only hold them into the winds that ripped the wave crests off. When a squall enveloped us and the rock disappeared from view, I finally got scared to the bone. I knew my boat had problems—the steering linkage had gone out the last time I had taken it out, and I prayed the strain of pulling another boat through the waves wouldn't break the mechanism. I felt a seed of panic sprout, not unlike the feeling I get in heavy air turbulence—"so this is how I'm going to go." I looked back at the "coasties" in tow. Waves were spilling into their boat and the poor bastards were bailing with their government-issued black shoes.

Eventually we gained the shoal waters, and the waves dampened out to a manageable 5 feet. Suddenly our boat was surrounded by fins. The panic attack subsided when we realized that they belonged to Bottle-nosed Dolphins and not to Tiger Sharks. As we neared home base, we slackened the tow line so that the land crew could pull them in. Our engine stalled out and the current quickly pulled us near the reef. The foremost thing in that situation is to "arrest further drifting," so we quickly threw out our anchors. They held too well in the coral and when we restarted our engine we had to cut them free. We could retrieve the anchors later, but now solid land was a priority and we tied up to the dock in short order. For our ordeal we received a

the direction of the wind drift, I thought I saw a signal flare through the curtains of mist, but I wasn't sure. Heading in that direction, we soon found the boatload of men. Soaked like wet wharf rats, they had scant survival gear, no raingear, no water, no anchor. We quickly threw them a line and began to tow them into sea swells building to about 10 feet.

measure of grog—brandy—our "exposure ration."
I headed directly to bed with the visage of La
Pérouse seared in my mind.

References

1. Grigg and Dollar (1980).
2. Armstrong (1983).
3. Grigg and Dollar (1980).
4. Beaglehole (1967:279).
5. Beaglehole (1967:604).
6. La Pérouse in E. H. Bryan Jr. (1942:177).
7. Amerson (1971).
8. *Pacific Commercial Advertiser* (1867).
9. Balazs (1979).
10. *The Friend* (1872:31).
11. Amerson (1971).
12. Amerson (1971).
13. Apple (1973).
14. Apple (1973).
15. For USCG personal recollections, see web page: www.radiojerry.com/frigate.htm.

8

The Hawaiian Monk Seal

Ua lilo i kai kuewa na kai kapu i ho'omalu ia.

"The protected seashores have become seashores for wanderers."
Cherished daughters have been led astray.

I thought there was no better profession than field biology as my Boston Whaler, throttle wide open, sliced through the calm, crystalline water of French Frigate Shoals. But if your study subjects are disappearing, it is no joyride to extinction. "Forty years ago, the monk seal population was about twice what it is today and it was low then," veteran biologist Karl W. Kenyon told me.[1] His 1972 paper, "Man versus the monk seal," awoke the scientific community to the endangered status of this Hawaiian native. Kenyon was training me in 1977, the first biotechnician in what has become a tag team of nonstop seal researchers. In my 3 months at the atoll, Kenyon taught me to recognize the age, sex, and scar patterns of the French Frigate Shoals seals, the last healthy population in the Leewards at that time.

Hawai'i has only two species of land mammals alive today that arrived under their own power. The Hawaiian Hoary Bat flew in from North America, settling in the forests of the main islands. The Hawaiian Monk Seal swam over from the Atlantic Ocean through the submerged Panamanian isthmus about 15 million years ago. Monk seals are pinnipeds, a group of marine mammals

that includes the walrus, sea lions, and seals. The monk seal has the most ancient lineage of all. It is characterized as a "living fossil" because its anatomy, physiology, and behavior have changed little over the years. They evolved in the North Atlantic Ocean and dispersed over the millennia to warm subtropical waters.

Three monk seal species formerly flourished in the waters of the Mediterranean, the Caribbean, and Hawai'i until hungry humans arrived. Because they evolved on remote oceanic islands and isolated sea coasts with no land predators, monk seals had not developed escape strategies. The land was a safe haven from sharks, their main predators. Sailors found the genetically tame seal reluctant to flee into the water and needed only a club to obtain fresh provisions. Christopher Columbus discovered the Caribbean Monk Seal on his second voyage to the New World. He recorded killing eight monk seals near Haiti. (He may even have eaten some Mediterranean Monk Seal steaks before he left Europe!) Five centuries later, the Caribbean Monk Seal is extinct and the Mediterranean Monk Seal numbers fewer than five hundred.

Although the monk seal resides in the Tropics, they are wrapped in a layer of blubber and covered with short, thick brown fur like most sea lions and seals. Male monk seals measure around 7 feet from nose to tail and weigh 380 pounds, females weigh 600 pounds, and both may live over 20 years in the wild.[2] The Hawaiian word for the seal is ʻīlioholoikauaua (dog running in the surf), but this is a modern name, because the seal does not appear in old chants or legends. It seems possible that seals were among the first species hunted to local extirpation by the early Hawaiians, but none of their bones has been found in Hawaiian archaeological sites in the main islands. The monk seals flourished in the northwestern islands until their discovery by Europeans. Sealing expeditions in the early nineteenth century almost eliminated them. Shipwrecked sailors, feather hunters, guano miners, and fishermen also took their toll until the seals were few and far between. In 1824, the sealing brig *Aiona* was thought to have killed the last seal. But seals foraging at sea escaped the decimation and returned to the islands. In 1859, the U.S. Navy bark *Gambia* took on board "240 bbl's., seal oil, 1,500 skins." Kenyon doubted the accuracy of the *Gambia* report. He did not think 1,500 seals could be taken during a 103-day cruise so soon after the slaughter of the population in 1824. During fact checking, he discovered that the original logs of the *Gambia* expedition were lost in the 1906 San Francisco earthquake and fire and therefore they could not be consulted.[3]

Nevertheless, seals had virtually disappeared. Max Schlemmer, lessee of Laysan Island at the turn of the century, could only kill seven seals during the 15-year period he lived there. In 1911, no seals were seen by biologists during 6 weeks on Laysan Island. But under protection in the Bird Reservation, enforced by U.S. Navy patrols, the seal population slowly recovered. In 1958, the seal population, counted on aerial and beach surveys in the Leewards, numbered 1,206.[4] The at-sea population is about three times the land count total, so the entire population probably never climbed higher than four thousand animals before it began to dip again.

Seal numbers were on the rise at French Frigate Shoals in the late 1970s. It was initially thought that the local population was recovering from human disturbance during the war years. The French Frigate Shoals population grew from a low of 35 in 1957 to 223 in 1977 and later climbed to 744 in 1983. However, in the far western section of the archipelago, numbers of seals were declining, and the species was declared endangered in 1976. The population rallied somewhat during the 1980s, but by the early 1990s, there were just over 1,500 Hawaiian monk seals, and the population was declining at about 5 percent a year. Luckily the slide halted around 1993, and recent counts show that the population has stabilized at around 1,350 seals, though the population at French Frigate Shoals is still in trouble.[5]

No one knows all the reasons for the overall decline, but several factors may operate in concert. Kenyon initially thought that "glass-ball fever" played a role on inhabited islands. Beach walkers at Tern, Midway, and Kure would occasionally find beautiful, hand-blown, glass fishnet floats in vari-

Hawaiian Monk Seal swimming off Tern Island.

Monk seal resting with nose in the sand.

favorite. The Tern Island Coast Guard station lasted until July 1979, when the loran system was replaced by satellite navigation. Tern Island was left for the birds, sea turtles, monk seals, and a few biologists. The seals returned almost immediately. In 1980, up to twenty-two seals used the beaches. On 5 March 1985, 181 seals were using the beaches, and the first pup was finally born there in 1988.

When asked why they are called monk seals, Kenyon replied, "because the people who study them live like monks." That was certainly true for me while stationed on Tern Island in 1977. But things change, and sometimes for the better. In 1990, almost 10 years after the Coast Guard left, I visited Tern Island and was gratified to see the coastline crowded with seals, and the swimming beach where the lonely men had swum

ous sizes and colors. With little else to do on-island and a fever-pitch desire to possess them, some men raced to the beach before sunrise to look for newly arrived glass balls. Seals, inadvertently startled awake from a deep slumber, eventually abandoned safe beaches for unsafe areas, sometimes near shark-infested waters.

Human disturbance had a lasting effect on the survival of pups. Kenyon reported, "Monk seals rarely haul out on inhabited Tern Island and there are no reports in recent years that pups have been born there."[6] As Kenyon and I walked the Tern Island beaches in 1977, we noted that the Coast Guardsmen and their dogs repeatedly chased the seals lying on the beaches into the surf. Only five juvenile seals consistently used Tern Island as a haul-out area. "Scarface," a little female, was my

Hawaiian Monk Seal pups and females, Gin Island, 1977. The small pup was born moments before the photo was taken.

was sometimes graced with the presence of top-less nymphs. (That's how the National Marine Fisheries Service's acronym [NMFS] is pronounced, and it is NMFS biologists, mostly women, that study the monk seal in the field.) On Tern Island, things had improved for monk seals and monastic researchers alike. Research in field camps stationed along the length of the Northwestern Hawaiian Islands continues to this day, with ongoing in-depth studies of seal reproductive behavior. But who is studying whom? I recall spending a lot of time studying my colleagues. Odd gender ratios and tropical nights make for some strained situations, especially if you are unlucky (lucky?). Still, having anything other than an all-American, 100 percent male camp is a breath of fresh air out here.

In 1977, the atmosphere was pure testosterone. A hefty bos'n's mate with a 6-inch knife scar on his belly (said to eat raw liver for breakfast) tried to pick a fight with me. "Boats" had sprained his ankle while he was playing volleyball. After several Demerol, an ice pack, and a six-pack, the "Lord of the Flies" was on the rampage. My cheap red sneakers that had faded to pink probably provoked his drunken ire. He called me a "fairy tern" and dared me to take a swing at him. But the fairy tern, aka White Tern, is one of my favorites, and the insult easily passed. I wonder what he would say to the researchers living on East Island who now wear panty hose to deter bird ticks?

A more likely reason seals are named after monks is because of the rolls of flesh that gather like a cowl around their neck. Plus, the relatively solitary existence they lead, rarely lying in contact together as do their closest relatives, the Elephant Seals, lends credence to their name. As the sun

sets and the coolness of the evening approaches, monk seals either move inland to lie among the vegetation, protected from the wind, or leave the island to feed on the creatures in the atoll waters. They eat various reef fishes including Conger and Moray Eels, as well as octopus and lobsters. Monk seals, feeding in the inshore coral reefs, stay submerged for about 10 minutes, enough time to successfully obtain food.[7] Seals also spend a portion of the year at sea and must be able to feed in waters of great depth. Dive recorders placed on monk seals' rear flippers have documented their descent to over 300 feet, but their deep-sea diet remains unknown.

When seals return to land after a long sea venture, their fur is sometimes green with algae. For days on end, they lie about in the hot sun, their fur bleaching. As the sun rises, they get too warm and move to the wet sand to cool off. To keep the numerous flies from disturbing their fitful sleep, they may bury their muzzle in the sand, sneezing frequently to dislodge sand, flies, and

Karl W. Kenyon censusing monk seals. In 1977, approaching seals was an accepted technique. Today, surveyors make special efforts to remain unseen.

nasal mites. Eventually driven by heat to the surf, these seals will alternately swim and sleep. Pregnant females haul out on a beach near shallow water to give birth, usually on the island where they were born. Seals tend to remain near their natal island, forming rather discrete populations with limited interchange. Pupping occurs from late December to mid-August, with the peak coming between March and May. This is a leisurely pupping season compared with that of Arctic seals, which must pup as promptly as possible, usually during 3 weeks of favorable weather. Female monk seals give birth to one jet black pup, born fully haired with open eyes and a ready appetite. The pup resembles a sugar-coated pastry when it rolls

in the sand after a swimming session. During the nursing period, the female does not leave the pup and so cannot feed herself. As she fasts and converts her stored fat into milk, she grows slim, while her pup balloons into a silver blimp. Weaning occurs around 36 days after birth or when the mother has run out of milk. Independent pups are silver gray and must learn to feed themselves. Until this is mastered, they live off their baby blubber. Young pups relax so fully in the surf that they can appear dead. I once pulled a "dead" pup from the surf only to have it wake and look at me in mild alarm.

We were very cautious when we passed by a female and pup on our surveys, because the

Male and female monk seals sparring on Laysan Island.

nursing cycle can be disrupted easily. When the mother is disturbed, the pup receives less milk and may be weaned with insufficient fat to survive long enough to learn to feed on its own. Yet, sometimes, just being in the seal's environment is disturbing. Once while censusing seals at Trig Island in French Frigate Shoals, I had to swim across a channel to survey a small islet. Suddenly, I was confronted by an angry mother seal. She charged at me, a wake roiling off her chest, her pink mouth wide open. She had me in her element and knew it. I could see every tooth in her head and anticipated a lunge at any second. Underwater, I struggled to get my footing in the sand as the irate seal passed behind me to cut off my escape. As she pulled in front of me again, I caught a toehold on the bottom and rocketed out of the water backward onto dry land. My close encounter with a defensive female was dangerous indeed. At worst, she could have drowned me. At best, a bite from her would have quickly become infected. Luckily, I was traveling with a Coast Guard medic who was standing in slack-jawed shock on the opposite side of the channel. I drew the female away so that he could quickly swim across. Fearing I'd lose my job for harassing an endangered species, I made him promise to tell no one.

On isles crowded with seals, confusion sometimes occurs and pups are swapped or even stolen by more aggressive females. The phenomenon of pup swapping, first observed in 1977 at French Frigate Shoals, reached such a high rate that twenty of the thirty-seven breeding females exchanged pups. Seven mothers exchanged pups once, eight swapped twice, four swapped three times, and one confused female swapped four times in 1985! Females rarely have enough milk for two pups and the larger, more active pup gets the lion's share while the younger one usually starves. The greatest mortality occurs after weaning, a factor that had prevented the species from full recovery until William Gilmartin of the NMFS began rescuing

orphaned and underweight female seal pups from French Frigate Shoals beginning in the mid-1980s.[8]

Charged with saving this "living fossil," Gilmartin sent more than ninety malnourished female pups to Honolulu for veterinary care and fattening, and then shuttled them up to Kure Atoll where seals were on the verge of local extinction. Young seals were kept in a large enclosure on the beach until they gained experience catching fish from the reef and reached a healthy weight. This active protection program, called "Headstart," helped the seal population increase in their western range. The first of the rehabilitated seals are now breeding on Kure. However, a recent downturn in seal reproduction at French Frigate Shoals has alarmed Gilmartin. It appears that the carrying capacity of the ecosystem of this atoll peaked around eight hundred seals in 1986. Since then, the population has decreased by about half. If the population falls any lower, seals may need to be moved back from atolls in the portion of the range where they are slowly increasing.

Seal population decline may also be due, in part, to a decrease in ocean productivity during the last decade or so. Gilmartin and the fishery scientists at NMFS cite evidence of a regional decline in fish, lobsters, and seals caused by a climatic regime called the Pacific decadal oscillation, further discussed in chapter 11. Another factor may be global warming. Certainly, change is constant, but water may be warming, bleaching corals in the process, and sea level may be rising. This rapid sea change is especially noticeable at Whale-Skate Island. In 1923, there were two vegetated islands, Whale and Skate. (A Pygmy Sperm Whale, very rare in Hawai'i, is the namesake for Whale Island, where a skeleton was collected by the Tanager Expedition. Skates and rays, which leap completely out of the water to display, may be the inspiration for the other island's name.)[9] These islets eventually merged into a single long island. When I camped there in 1977, the island had eight species of plants

Black pup nursing from mother on Laysan. Behind the female is a weaned pup from another female trying to ingratiate itself and steal milk.

refuge from large sharks on the prowl.[11] Adult female seals with nursing pups or weaned pups prefer shallow-water areas protected by reefs that exclude large sharks. Smaller Gray Reef Sharks use the same shallows as the seals, and nursing seals have attacked reef sharks stranded by the tide. However, larger Whitetip Reef and Galápagos Sharks are a threat to seals in deeper water.

Periodically, diseases cause seal populations to decline. Biologists rushed to Laysan Island in 1978, when an estimated fifty seals died. Some recovered carcasses were autopsied by pathologists, who discovered the cause of death to be an outbreak of ciguatera poisoning. Ciguatera is caused by a toxin-producing microscopic dinoflagellate, algae that grow on certain brown and red seaweeds that colonize disturbed coral surfaces.[12] The toxins build up in herbivorous fish, which, when eaten, pass on the toxins to carnivores. As top food chain predators, seals from Laysan may have eaten eels that had concentrated the toxins of reef fish. To test this theory, Bill Gilmartin fed some eels to captive Elephant Seals, which quickly succumbed to the neurotoxin. However, the Elephant Seals had not evolved in a habitat where ciguatoxins occur naturally, and it is likely that monk seals probably have some resistance to the toxins. Biologists speculate that harbor dredging at Midway Atoll may have created conditions that provoked this lethal outbreak. In 1992, another outbreak occurred and killed sixteen monk seal pups that had been released at Midway.

Man-made marine pollution, in the form of oil, fishing hooks, and nets, is increasing and may be contributing to the seals' overall declining health. In 1993, an oil slick hit the beaches of Laysan Island

including Beach Morning Glory vines, beachgrass, and Beach *Naupaka*. By the mid-1990s, Whale-Skate Island had begun eroding and it is now completely under water.

Seal predation may also be increasing. Throughout their range, seals must contend with Tiger Sharks. In 1977, biologists from the Waikīkī Aquarium caught twenty-three large Tiger Sharks at French Frigate Shoals (an estimated five hundred live in the waters outside the shoals) and found that three sharks had monk seal remains in their stomachs.[10] In the island surveys, about 12 percent of the seals showed evidence of shark bites, though it is not clear what degree of mortality results from such bites. That they can survive a shark attack at all is incredible testimony to the power of healing that these seals possess. One way seals may avoid sharks is to retreat into underwater caves. Scuba divers at La Pérouse Pinnacle discovered six caves that served as monk seal hideouts. They saw seals breathing from bubbles of fishy smelling air contained at the ceiling of the crevices. These submarine lairs may serve as a temporary during the beginning of pupping season and oiled at least fifteen monk seals. Because the process of cleaning oiled seals can be more damaging and stressful than the oil itself, NMFS decided to allow nature to take its own course. It appeared to work that time. Seals are also attracted to fishing gear and have been found wrapped in nets and lines, sometimes appearing onshore with hooks imbedded in their flesh. Seals entangled in fishing gear prompted NMFS in 1990 to invoke an emergency ban of long-line fishing within waters less than 50 miles from shore. Nevertheless, literally thousands of hooks are set outside the 50-mile boundary, and nothing prevents

Monk seals and sea turtles on East Island, French Frigate Shoals.

the seals from visiting these potentially deadly sets. In late 1999, a federal judge closed all waters within 200 miles of Hawai'i to long-lining until an environmental impact statement could evaluate the harm to nontarget species.

Seals may also be affected by pollution-caused changes to their sex hormones. On some islands, aggressive male monk seals outnumber females three to one. Without a single dominant male to control the mob, the males gang up on females. Males attempting to mount an estrous female grab her back skin with their teeth and rip her open. Bleeding back wounds attract sharks and an unknown, but possibly substantial amount of female mortality results. Mobbing behavior has been observed at French Frigate Shoals and Laysan, and badly scarred females can be found on all islands. To deter male aggression, nine males were taken 600 miles south to Johnston Atoll and released in 1984. Another twenty males were shipped off and exiled to the main Hawaiian Islands in 1994. They all disappeared. "Mobbers" were also identi-

fied and injected with a drug designed to counter their testosterone output. These approaches did not seem to work too well, because attacks were still recorded. Infanticide by males is also a problem. In 1990, a large male killed four pups on Whale-Skate Island at French Frigate Shoals. His attempted killing of a fifth pup sealed his fate. The NMFS biologists decided to destroy him by lethal injection before he had a chance to kill again. Endangered species management by capital punishment may seem a dangerous way to "save" seals, but this species needs every female and pup if it is to survive into the twenty-first century.

References

1. K. W. Kenyon, 1977, personal communication.
2. Kenyon and Rice (1959).
3. Kenyon and Rice (1959).
4. Rice (1960).

5. National Marine Fisheries Service monk seal biologist Jason Baker, June 1999, personal communication.
6. Kenyon (1972:688).
7. Rauzon and Kenyon (1982).
8. Winning (1998).
9. Wetmore in Olson (1996).
10. Taylor and Naftel (1978).
11. Taylor and Naftel (1977).
12. Withers et al. (1980).

The Green Sea Turtle

Pūhā hewa ka honu i ka lā makani.

"The turtle breathes at the wrong moment on a windy day."
A person says the wrong thing at the wrong time and suffers as a result.

Save the sea turtle: E ho'opakele i na honu o ke kai!

As I snorkeled off Waikīkī, a Green Sea Turtle the size of a suitcase grazing on the algae growing on a dead coral head tolerated my proximity. I noticed the short tail that indicated this turtle was female. Her silted shell obscured her tortoiseshell pattern, so she seemed to appear and disappear as she drifted in the murky water. She was an ambassador from French Frigate Shoals, the sea turtle capital of Hawai'i. Green Sea Turtles had nested on isolated beaches throughout the main islands until the mid-1950s, and although turtles are present near shore around all the islands, 90 percent of Hawai'i's Green Sea Turtles nest at just two islets in French Frigate Shoals.

Well before the Hawaiian Islands ever exploded above the ocean surface, sea turtles glided over them. Later, these ancient and elegant animals hauled out on the first Hawaiian beaches some 100 million years ago to lay round, gooey eggs before crawling back into the warm sea. Five species of marine turtles visit Hawai'i. The Hawksbill Sea Turtle is a rare and endangered species, nesting in small numbers on the black-sand beaches of the Big Island of Hawai'i and on Maui. Leatherback Sea Turtles have been seen in the pelagic zone surrounding Hawai'i but do not nest in the state. Loggerhead and Pacific Ridley Sea Turtles, officially listed as threatened species, only occur as rare migrants.

The Green Sea Turtle or *honu* is the best-known marine reptile in Hawai'i and throughout the world. It is also the most valuable reptile in the world because of its economic importance to coastal dwellers of the Tropics. The meat, the greenish fat from inside the top of the shell, and the cartilaginous "calipee," found next to the lower shell, are made into soup. The shells are dried and polished for souvenirs, and the demand for turtle eggs has caused the species to decline worldwide.

Green turtles are recovering from wholesale harvest during the last 150 years in Hawai'i. Expe-

Sea turtles and a monk seal harvested on Laysan Island. (Hawaii State Archives photo)

ditions to the Northwestern Hawaiian Islands exploited once-numerous marine resources, including seals, sea turtles, sharks, and sea cucumbers (bêche-de-mer). The fishing schooner *Ada* took over 410 turtles in 1882.[1] Over 60 sea turtles were collected in 1 day from Laysan, and another 43 were taken in a few more days. The crew found a board painted with an appeal not to take turtles away. They repainted the sign before departing with 103 sea turtles. The ship returned several months later and took 26 more. An account from the log of the Rothschild Expedition recorded that East Island at French Frigate Shoals was formerly known as Turtle Island because hundreds of them basked on the beach while ten times as many bobbed in the surf.[2]

There is a real sport on these sand islets. Here the turtles take their midday siesta by the thousand.

You may, with your companions, each select a big turtle on the beach and, if you are quiet and quick, stand on his back, and the race to the water begins. If the turtle scents you first, he digs his flippers in the sand and sends a blinding shower of sand into your face. If you are expert in the water, and know how to handle a turtle, you do not have to dismount—just keep your left knee on your turtle's back, your right hand under the shell back of his head, and use your right foot in the water as a rudder to guide. You can make your turtle swim in the direction you wish him to go; moreover, if he attempts to dive, just lean back and pull upward on his shell with your right hand . . . .Then, too on the shoals there is the sport of digging in the sand for turtle eggs, a hundred round, soft-shelled eggs in a nest a foot or so in the sand.[3]

Captain sitting on sea turtle, Laysan Island. (Bishop Museum photo)

upper dry beach. After 50 to 70 days, depending on the temperature in the nests, the eggs hatch and the young turtles scramble to the surf. Along the way, some fall prey to ghost crabs. In the ocean, they must evade predators like sharks and jacks while presumably hunting zooplankton, their one stint of carnivory, because Green Sea Turtles are herbivores. After hatching, sea turtles are rarely seen until they are about 15 inches in length. Where they go during this growth period is still a mystery of the sea. But only about 1 percent of the turtles survive to breeding size.

In 1959, a fishing crew took 25 percent of the turtles at French Frigate Shoals. The profligate slaughter caught the government's eye and prompted some protection.

George Balazs of the National Marine Fisheries Service has been studying Green Sea Turtles at French Frigate Shoals since 1973. The foremost expert in Hawai'i on this species, his rallying cry: *E ho'opakele i na honu o ke kai*—"Rescue the turtles of the sea!" has been heard loud and clear. Numbers of Hawaiian Green Sea Turtles have increased under strict law enforcement since 1978 when they were listed as threatened under the U.S. Endangered Species Act. When I was at Tern Island in 1977 there were no turtle nests. Twenty years later, there were over 700 turtle nest pits! On East Island, when Balazs began his research, there were 69 nestings. In 1997 there were 504. During his long-term research he has tagged 2,000 breeding females as well as 3,900 others in Hawaiian waters. These tagged turtles represent most of the population.[4]

Sea turtles take an average of 25 years before they begin breeding. Nesting begins in May and continues through August, during which time a female lays 100–150 eggs in a sand pit dug into the

Adult turtles often come ashore missing rear flippers and bite-sized portions of their carapaces. A study of Tiger Sharks found that the relatively slow-moving sea turtles are a substantial portion of their diet. Scientists fishing in 60 feet of water off Shark Island at French Frigate Shoals caught twelve sharks on ten hooks. Two 7-foot Galápagos Sharks got caught and attracted two Tiger Sharks, which ate the Galápagos sharks in a couple of bites, hooking themselves in the process.

Green SeaTurtle missing a rear flipper lost to shark bite.

The largest Tiger Shark, a magnificent monster over 14 feet long, had to be dispatched with four shots from a carbine rifle. In its stomach were six whole sea turtles, a shark's head, a Spinner Dolphin skull, one monk seal flipper, and the remains of Wedge-tailed Shearwaters, Eagle Rays, and Spiny Lobsters. But sharks and sea turtles have evolved together for millions of years, so turtles have obviously found ways to survive in this dangerous environment.

The Hawaiian population of sea turtles is the only one to sleep out of the water. Basking is common, especially during breeding season, and may be a behavior that evolved as a protective measure against Tiger Shark predation as well as a method of raising their body temperature at locations where sea temperatures are cool. They usually bask on warm, sandy beaches but have even been observed basking on a shipwreck and on a huge tangle of fishnets at Pearl and Hermes Reef.[5] Land basking has allowed Balazs excellent opportunities to study these creatures. By tagging turtles with numbered bands and gluing radio transmitters to their shells, Balazs has discovered their migration patterns. Three turtles were fitted with transmitters in 1991: two migrated from French Frigate Shoals to Kāne'ohe Bay on O'ahu, and one reached Johnston Atoll 400 miles to the south. Swimming south of the islands over deep water, they travel fast, averaging 1 mile per hour to return to their preferred foraging areas.[6]

Unfortunately, although the threatened Green Sea Turtle is recovering from overharvest, they have begun to suffer from a new disease. First spotted in Florida, it has spread to all oceans and increased alarmingly in Florida and Hawai'i during the 1980s. Turtles, especially large juveniles, develop noncancerous tumors around their neck, flippers, eyes, mouth, and internal organs. Called fibropapillomatosis, it is not fatal in and of itself, but the tumors block the throat, eventually causing starvation and increased vulnerability to entan-

Green Sea Turtles mating in the surf at Tern Island, French Friate Shoals.

glement in fishing gear. As many as 12 percent of the nesting females at French Frigate Shoals have tumors, but over 50 percent of the turtles examined in Kāne'ohe Bay, O'ahu, exhibit them.[7]

The exact cause remains unknown, but a herpes virus has been detected in the tumors.[8] The turtles' immune system may be compromised by environmental factors such as pollution, water temperature changes, or increased ultraviolet light exposure caused by the thinning of the atmospheric ozone layer. Scientists have attempted to treat the sea turtles with antibiotics and surgically remove the external tumors when possible; however, the tumors do regenerate and internal ones cannot be removed. Development of a vaccine is unlikely, so researchers are hoping to eliminate the possible causes in the ocean.

The large female turtle I saw off Waikīkī has probably been returning to French Frigate Shoals every few years to breed and will probably continue to do so for several more decades. In spite of the threat of a new disease, her breed are survivors since before the days of the dinosaurs.

References

1. Balazs (1980).
2. Walker (1909).
3. Hudson (1911:340–341).
4. Rillero (1999).
5. Kam (1984).
6. Rillero (1999).
7. Barrett (1996).
8. Rillero (1999).

10

Albatross—Lord of the Wind

'Au i ke kai me he manu ala.

"Cross the sea as a bird."

The Northwestern Hawaiian Islands harbor more than 90 percent of the total Hawaiian Archipelago's seabird population. Two species of albatrosses, two shearwaters, two petrels, one storm-petrel, one frigatebird, one tropicbird, three boobies,

Laysan Albatross in flight.

and six tern species have major breeding populations in the Leewards. Approximately 14 million seabirds live there, including 5.4 million breeding pairs of eighteen species. (French Frigate Shoals is the only island where all eighteen species nest.) The Northwestern Hawaiian Islands populations of Laysan and Black-footed Albatrosses, Christmas Shearwaters, Bonin and Bulwer's Petrels, andTristram's Storm-Petrels are the world's largest.[1] These vast seafowl rookeries are among the most important and diverse in the world, partly because they are the most protected and least disturbed colonies.

With a wingspread of 8 feet, the impressive albatrosses best represent the Northwestern Hawaiian Islands. The god Kāne can take the form of the albatross, called *mōlī* in Hawaiian, which is also the word for bone tattoo needle. Albatross wing bones were used as implements to inscribe human skin with designs. The *mana* of the bird was also invoked: to sail expertly on the winds. The sea breezes lift them up and they bank high above the sea by vertically tilting their saber-shaped wings. The birds slice through the air and dip down to the water, perhaps scoring a wake with their wing tip. An albatross levels out over the waves without a wing beat while the air, pushed by the waves, lifts them up. Again and again, as long as the wind blows, the albatross repeats this energy-saving flight, called "dynamic soaring." An albatross can sail for days and may fly millions of miles during its 50-year lifetime.

Albatrosses generally forage north of the Hawaiian Islands, ranging into the Bering Sea and east to the California coast. Severe weather is spawned in the North Pacific, and albatrosses are

A Black-footed Albatross poses for a portrait.

at their best in it. In gale-force winds they glide with breakneck speed; indeed, it is safer to be airborne than to sit on the foaming sea. The windless regions to the south present a barrier to albatrosses. Without wind, the birds become becalmed. They need some breeze to lift their wings as they run on the sea surface to take off. The belt of stagnant air around the equator, called the doldrums, caught sailing vessels for weeks without movement. Animals died of thirst in the relentless heat and were thrown overboard. Bloated carcasses floated at the surface until the sharks found them, and the sailors who saw the carnage called that area the "horse latitudes."

Albatrosses are the grandest members of the tubenose family, the procellariids, which also includes shearwaters and petrels. Nicknamed "tubenoses" because of their ornately structured beak, members of this seabird family have unique glands in their head to remove salt from their body.

Concentrated droplets of saline exude from the tubes atop their beak and flow down a channel to drip from the nail-sharp tip. Tubenoses are also able to smell, an ability rarely found in birds. As the wind flows through their nostrils, they can sense odors that may indicate a dead marine mammal is floating upwind.

Around one million Laysan Albatrosses and 100,000 Black-footed Albatrosses return to nest in the Northwestern Hawaiian Islands every November.[2] You can almost set your clock by their return. The black-foots arrive during the last week of October and for several years running laid their first egg on 14 November. Laysan Albatrosses arrive in the first week of November, and the first eggs appear around 20 November. When the caterwauling, cake-walking albatrosses first encounter each other, mature birds seek out their old mates and young birds search for a new dance partner. They experiment with many individuals before they find a perfect partner. Once located, their pair-bond is lifelong.[3]

Albatrosses have some of the most complex mating behaviors of any bird, and the courtship behavior lasts all season long. As long as the birds are on shore, they are reinforcing their pair-bond through a repeated repertoire of sound and action. Whines, whinnies, and songs of the courting albatrosses are the language of love on these romantic isles. When pairs meet, they exchange a bowing and mewing recognition, followed by a face-to-face stance in which birds clamor while rubbing bills. One bird reaches its head back to nibble its wing, a preening behavior adapted into the dance. The other bird stands at rapt attention, facing its partner and rapidly "clappering" its bill. Suddenly one thrusts its bill skyward and utters a drawn-out moan that ends on a high note. After a few more solo "sky-calls," both begin clacking like castanets. Then, in perfect coordination, they simultaneously sky-call and groan in exaltation.[4] Laysan Albatrosses execute a more elegant version of the

Black-footed Albatrosses flock with one Laysan Albatross.

egg for about 4 days before relinquishing incubation duties to the male. His shift is quite a bit longer. The female goes to sea for an average of 22 days to feed to make up her lost weight before she returns. During the 65-day incubation period, both parents take other extended forays, ranging far across the Pacific for food, including squid, flying fish eggs, and deep-sea fish. In 1998, researchers placed sophisticated satellite tracking devices on albatrosses to record their foraging flights. One Laysan Albatross flew over 2,000 miles to the Aleutians to feed. It returned to provision its chick with the nutrient-rich oil from its stomach, then departed the next day for the same Alaskan site![5]

It is July before the parents complete their chick-rearing duties for the year. Fed on a diet of pre-digested flying fish eggs, squid, and red stomach oils, albatross chicks grow steadily in bulk and weight. But by mid-June, their parents feed them less and less frequently until they cease visiting the island entirely and depart for the North Pacific, the Gulf of Alaska, and the Bering Sea. As the summer solstice approaches, the young birds are the size of adults but exceed the adults' weight by almost a third. The soon-to-be fledglings wait in vain at their nest site for their parents to return with a meal. As they wait, they slim down from overweight butterballs to svelte, if awkward, adolescents.

The sun beats down and heat waves dance off the blinding white sand of East Island, French

dance than the aggressive style of the husky black-foots. Occasionally, signals get crossed and hybrids between the two species result. When they reach adulthood, hybrids dance like Black-footed Albatrosses but flock with Laysan Albatrosses, so they never get the dance right for either species.

The result of this dance is a single white egg, about 5 inches long, that both parents take turns tending. The female lays and remains on the

Laysan Albatrosses courtship dancing.

The complex courtship dance of the albatross is a series of rit-ualized movements set to the music of hoots, whines, and whistles. A pair performs the "wing-tuck."

Black-footed Albatross courtship dancing.

Frigate Shoals. Several hundred young albatrosses, both Black-footed and Laysan, are ready to fly to sea. They have weathered late winter storms and summer heat waves during their first 150 days of life. Now, their greatest challenge is to become air-borne. They spend the summer days panting in the heat and exercising in cool weather. Each gust of wind, especially when accompanied by rain, elicits the same response from thousands of alba-trosses. When a squall sweeps across the island, the accompanying winds stimulate all the fledg-lings to open their wings to the breeze and prac-tice in unison. They all jump up and down, flap-ping their narrow wings. Their flight feathers are stiff, and tufts of down are wiped off during this exercise and carried away by the trade winds. The chicks are well-feathered, but vestiges of down still cling to their heads and necks, giving them the look of Elvis impersonators with their various "ducktails" and sideburns.

Many fledglings gain experience by flying short distances and crash landing. Over and over, they practice until they develop enough coordina-tion to control their wings in the trade winds and escape the island altogether. Individuals who live near the shore have a clear runway to practice on. Free from obstacles, they run along the beach and fly for the first time. But out there, unlike in the interior of the island, it is often a one-way trip. A fledgling makes a running start from the sand and hits the water, flapping its wings while running on the surface with webbed feet. The youngster gets airborne briefly, but its new flight muscles tire quickly. Wing tips drag through the water, and the still-downy plumage becomes wet. Weighed down by water, the albatross pauses to rest and perhaps consider the novel medium that appears to offer

Black-footed Albatross feeding chick a slurry of red stomach oils.

The Last of the Mohicans—a fledgling tries its wings.

Tiger shark eating a Laysan Albatross fledgling. The shark protects its eyes with a nictitating membrane. It bumps the albatross, then swamps the bird in passing. Sure of its mark, the shark turns and eats the albatross.

an escape but ends up confining them.

As the bird floats on the surface, an offshore current pulls it farther from the island. The bird thrusts its wings out and runs again on the surface, but each time it tires sooner. As the exhausted bird drifts near the reef, an underwater shadow slowly approaches. Suddenly a 14-foot Tiger Shark erupts from the water. Its snout reaches 2 feet in the air to the height of its prey. Closing the nictitating membranes of its eyes, the shark brushes against the unaware albatross. A head bump and backwash of the passing shark swamps the bird. The shark is now sure of its mark. It turns instantly, and with one mouthful chops the bird into oblivion. In its wake, a "footprint" of albatross stomach oil calms the surface. An expression of death floats like a temporary grave marker, and the daily death toll can be counted on the water until the oil and blood disperse into the great ocean.

East Island, French Frigate Shoals, is one site where a breach in the reef allows large Tiger Sharks close access to the shore where birds enter the water. Usually, Tiger Sharks reside in the deeper waters outside the barrier reefs of the Northwestern Hawaiian Islands. They enter lagoon waters at night and usually leave by daylight. But during albatross fledgling time at specific sites off all the islands in the northwestern

A pregnant female Tiger Shark about 14 feet long, caught as part of a study of food habits in 1977.

References

1. Harrison (1990).
2. U.S. Fish and Wildlife Service biologist Beth Flint, 2000, personal communication.
3. Rice and Kenyon (1961).
4. Rice and Kenyon (1961).
5. *The Honolulu Advertiser* (1998).

archipelago, they reliably appear. Certain individual Tiger Sharks learn of the ephemeral bounty available to them, and they generally appear in the morning hours during a brief period in July to take their fill. Small sharks are scared away from the feeding grounds by the larger Tiger Sharks, so only the largest sharks take advantage of the naive birds. The albatross' moment of death ends an 8-month period of labor by the adults, their annual output devoured in an instant. Albatrosses are long-lived birds, so they will have other chances to fledge a chick successfully. Next year, the chick may survive.

11

The Fabric of the Sea

Ka iʻa ʻimi i ka maona, na ka manu e haʻi mai.

"The fish sought for in the ocean, whose presence is revealed by birds."
Tuna below the surface is revealed by the birds.

The giant Jack or *ulua* eyeballed me at close range, seriously contemplating my exposed flesh as I snorkeled off La Pérouse Pinnacle. It is a hungry ocean, so I quickly swam to the boat to keep from joining the local food chain. As we motored away, the *ulua* struck the prop of our outboard engine, hoping it was the flash of a fish—a jumping jack flash. I'm grateful I am *almost* big enough to be exempt from predation. But when a 14-foot Tiger Shark swims under your rubber boat, you pray that you are really living in a parallel universe and part of a land food web where you're almost immune to predation (except from your fellow man).

Tiger Sharks are the top food chain predators: "the bite stops here." They eat other predators such as Gray Reef Sharks and Galápagos Sharks, but are mainly opportunistic feeders. Their diet consists of 45 percent reef fish, 24 percent sharks, 20 percent seabirds, 4 percent tunas, 4 percent lobsters, 2 percent sea turtles, and 1 percent monk seals.[1] They rarely add humans to their diet. When they do, or after a particularly predaceous episode occurs against monk seal pups or sea turtles with radio transmitters, there is a call to remove sharks from the ecosystem. Sharks that have attacked people are usually long gone when the authorities come looking for revenge. A recent study on Oʻahu found that Tiger Sharks have tremendous ranges and may travel over 40 miles in a night. Sharks innocent of manslaughter no doubt pay for the sins of their kin, but they are only following the law of the sea. Eat or be eaten!

When predators are removed, the food web adjusts in an unpredictable manner. One experiment, albeit unplanned, has occurred in Hawaiʻi. The wholesale removal of top food chain predators via fishing has been going on for many years. The removal of large sharks, tuna, and billfish by the fishing industry has increased throughout the Pacific since 1948 and in Hawaiʻi during the last two decades.[2] As a result, these top predators have declined to 20–50 percent of their 1950s abundance levels. Their absence creates a "competitive release," and food items usually consumed by top predators become available to secondary predators like seabirds and smaller fish.[3]

There has been a noticeable increase in second-level predators in Hawaiian waters. Fast-growing, short-lived animals such as *mahimahi*, squid, flying fish, and small tuna respond quickly

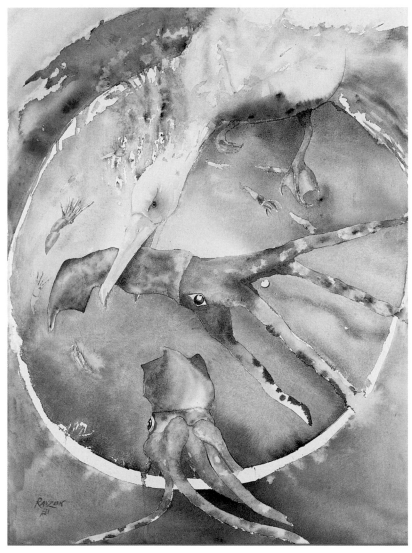

Laysan Albatross seizing a squid.

and prey takes a strange twist with seabirds. Species like boobies rely on large fish to chase small fish to the surface where they become available to the birds. If the tuna fishery around Hawai'i removed too many tuna, then there might be an insufficient number of predators to drive prey to the surface. Then, in spite of an abundance of food made available though an increase in ocean productivity or "competitive release," prey species would remain unavailable to the birds. This line of thinking only opens more questions and speculation. What is the critical number of predators needed to drive prey to the surface? Could boobies point the way to better tuna management? "More research is needed," is the mantra.

Sometimes, all the research in the world can't stop economic forces. For example, National Marine Fisheries Service (NMFS) researchers were studying lobsters to determine the feasibility of harvesting, without exploiting, them. Hawaiian Spiny Lobsters or "bugs" are thought to be endemic to the Hawaiian Archipelago, Wake, and Johnston Atolls. Scientists discovered that the lobster population at Necker Island was estimated to be the highest in the Hawaiian Island chain. The vast underwater platform of the island (especially within the 20-fathom line) is full of *puka*s (holes) where lobsters reside. Juveniles hang out in big holes, dormitories in the reef, and adults room alone, hiding from their daytime predators such as sharks and jacks. By night, lobsters scour the reef for benthic,

to the "competitive release," whereas long-lived predators like seabirds respond more slowly. Fishery experts suggest that flying fish are about 25 percent more abundant now than they were 50 years ago. *Mahimahi* (the Dolphin fish), Red-footed Boobies, and Wedge-tailed Shearwaters are about twice as abundant in Hawaiian waters.[4]

The intimate relationship between predator

Zooplankton.

or bottom-dwelling, organisms to eat, dead or alive. They were so common at Necker that, as a result of intraspecies competition, the average sizes of males and females were the smallest in the archipelago. Hawaiian Spiny Lobsters are also less fecund than other species in the genus. Other species produce from 310,000 to 525,000 eggs; local females carry fewer but larger eggs.[5]

Lobsters are of great importance in sport and commercial fisheries. By the early 1970s, catch rates declined in the main islands because of fishing pressures, but around that time, the lobster resource at Necker Island was accidentally discovered. Skip Naftel, the formidable captain of the fishing vessel *Easy Rider,* was setting deep-sea fish traps at Necker Island when a big storm carried his traps inshore. They lodged at the 20-fathom line

and when he retrieved the traps they were filled with "bugs." For years, the *Easy Rider* had easy pickings. From 1974 to 1977, two boats worked the banks and shoals, taking about 5,000 lobsters per trip. Naftel was able to keep the fisheries managers and the competition in the dark for several years. But eventually word leaked out. Bigger and better ships were constructed and began to work the region extensively during the late 1970s, at the peak of ocean productivity. Even with the competition, *Easy Rider* pulled in 650,000 lobsters in 1980 alone.[6]

During that period of exploitation, the percentage of harvested lobsters declined from 54 to 23 percent of a population estimated to average 67,766 legal-sized lobsters. The catch rate declined from fifteen lobsters per trap-night in 1974 (with a three to one male to female ratio) to two per trap-night in 1984.[7] Undersized lobsters and egg-carrying or "berried" females were released at the surface and exposed to predatory fish as they drifted down

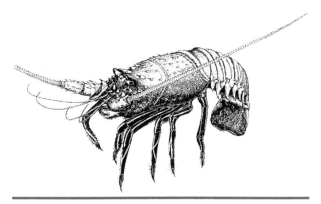

Spiny Lobster, Panulirus marginatus.

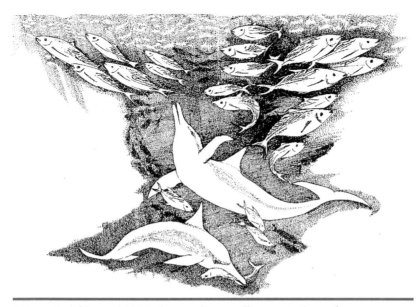

Spinner Dolphins and Skipjack Tuna chase smaller prey to the surface.

times more nutrients to the surface than usual in the mid-1980s. The ocean food web flourished and stimulated the bottom-up expansion of the food chain. When the Aleutian low slipped back to the northwest in the late 1980s, ocean mixing diminished, contributing to the population declines in fish, seals, and lobsters.[9]

This climatic phenomenon is called the Pacific decadal oscillation (PDO) because a review of past climatic data shows that the low-pressure system seems to shift with a near-decadal periodicity.[10] The fairly regular oscillation may be a response to an 11-year solar cycle of sunspots. The eruption of solar flares with a near decadal frequency increases heavy solar particles (atoms), which take approximately 6 months to reach our atmosphere. Solar particles reacting with the upper atmosphere may lead to higher atmospheric pressures, which may intensify and displace the Aleutian low-pressure system to the east. Recent oceanographic research has shown that Pacific decadal oscillation is part of a greater cycle defined as near-century oscillation on the order of about 70–100 years.[11]

Research is certainly needed to better understand the role of climate in ocean productivity. The infamous El Niño, until recently the province of marine biologists, meteorologists, and oceanographers, is now common knowledge, especially after the notorious 1997–1998 event. El Niño, the familiar Spanish term given to the Christ child, is used to describe a Christmastime shift in the flow of ocean waters along the west coast of South America. This periodic winter sea change is the result of the nutrient-rich cold water of the coastal Humboldt Current being replaced by eastward-flowing warm ocean water (which is nutrient poor) from the equa-

thousands of feet toward the safety of their dark holes in Necker Island. Released lobsters seldom survived and were lost to the breeding population. Also, because only males could be retained, the proportion of females in the catch increased and more individuals then were released to a waiting school of hungry fish. Under these conditions, the fishery declined drastically. Within 5 years, the lobster industry went bust. It seems that every commercial fishery has been managed to commercial extinction. Overexploitation rules; boat payments demand it.

However, although some scientists say overfishing destroyed the industry, others suggest climate change as the culprit. Researcher Jeffrey Polovina of NMFS discovered that ocean productivity, beginning in the late 1970s and continuing for a decade, was atypically high in the region.[8] The Aleutian low-pressure system, the dominant meteorological feature in the winter and spring in the North Pacific atmosphere, creates an eastward shift in winds that helps stir up ocean water. The extreme eastward position of the low brought five

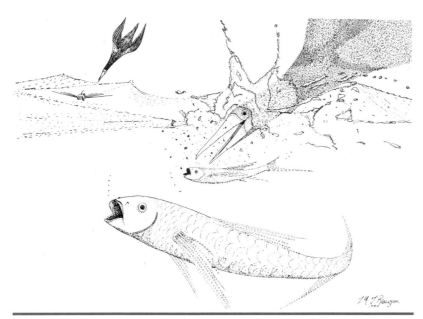

The Red-footed Booby depends on predatory tuna to drive flying fish to the surface, where they become available to these "plunge-divers."

ocean food chain. Just how did the cumulative seabird diet affect the potential for commercial fishing? USFWS researcher Craig S. Harrison tried to figure out how much and what the seabirds ate. I helped Harrison's cadre of biologists collect thousands of samples of bird "barf" in 1981. Grabbing and up-ending a seabird often caused the bird to vomit, thereby providing a nonlethal method of obtaining food samples. These boluses provided a window into the food web of the tropical Pacific Ocean.[12]

Because of their tameness, boobies were thought to be easy subjects to study. But biologists collecting regurgitations soon learned the truth. The birds can be aggressive when provoked, and removing them from a hand net requires speed, dexterity, and courage. We called them "pterodactyls" because their dagger-headed profile is reminiscent of the ancient leathery-winged reptiles, and their sharp-pointed, serrated beaks are lethal weapons to flying fish and biologists alike. The birds instinctively stab at the eyes and more than one biologist left the field bloodied but wiser to their ways.

It is hard to appreciate the business end of a booby when it is trying to slice you open, but all boobies have beautiful faces. Male Red-footed Boobies have orange skin surrounding their blue beaks, and females have pastel pink faces. The Brown Booby male has a dark bluish face, and the female's is yellow. The male Masked Booby's bill is chrome yellow and the female's is straw-colored, surrounded by a mask of indigo. Sexes of boobies can also be told apart by voice as well as by the subtle variations of skin and bill color. The male Masked Booby whistles, but the female honks. 'Ā is the Hawaiian name for all boobies

torial Pacific. El Niño events occur every 3 to 5 years, on average. There were three extreme events during the period from 1950 to 1980, and four major El Niños just since 1984. They can cause dramatically increased or decreased rainfall, which can lead directly to natural disasters such as floods or droughts. In Hawai'i, drought is the usual consequence, and its effects on the marine environment can be severe. The phenomenon deprives the seabirds of food because the ocean becomes too hot to nurture phytoplankton. The food web deteriorates and seabirds may forego breeding or starve in the worst cases. As these natural depletions subside, however, bird populations recover over time.

In the early 1980s, a "tripartite agreement" among the U.S. Fish and Wildlife Service (USFWS), the NMFS, and the State of Hawai`i allowed the creation of a program to assess the full range of marine resources in the Northwestern Hawaiian Islands. Governmental agencies were aware that millions of seabirds had a substantial impact on the and is probably derived from the drawn-out guttural *ahhhhhhh* that boobies sometimes make.

Boobies capture their prey by plunge-diving: dropping out of the sky headfirst, plunging under water, and grabbing a flying fish or squid. Different depths and locations for foraging allow the three species to avoid competition, in spite of the fact that Masked and Red-footed Boobies eat the same species of flying fish. The Red-footed Booby takes smaller prey, farther from land than the Masked Booby. Also red-foots sometimes snag flying fish on the wing. Masked Boobies are deep plunge-divers, feeding deeper in the water column than other seabird species; the Brown Booby feeds in shallow waters (often close to shore in O'ahu waters) and thus avoids competing with the other boobies for the limited food resources of the tropical seas.

Red-foots are the smallest species of booby. Nimble enough to build stick nests in shrubs, they cling to branches of Beach *Naupaka* with their webbed feet. As they perch on shrubs, their feet are conspicuously red year-round, although somewhat duller in the nonbreeding season. Other boobies nest on the ground. Brown Boobies build stick nests on flat islands and rocky promontories. They lay several eggs, but at most one will survive. Masked Boobies lay two eggs on bare sand; a patch of splattered excreta describes a circle around their nest. The strongest chick, usually the first hatched, will outcompete the smaller sibling until it dies from exposure or lack of food. Siblicide ensures that the fittest survives.

Boobies are siblicidal, but frigatebirds are kleptoparasitic: they steal for a living. Great Frigatebirds,

also known as man-o'-war birds, or *'iwa*, can tell if a booby has fed by detecting a bulge in its throat as well as by its labored flight. The expert fliers ambush boobies by upending them in flight as they return to the colony after a day of fishing. They attack the incoming boobies over the surf, force them to regurgitate their meal, and then catch the bolus of food before it hits the water. Frigatebirds rely on their long, hooked beaks and expert agility to steal prey from boobies, but they can also snatch flying fish on their own.

Some frigatebirds specialize in their prey, as biologist Laura Gill discovered on Tern Island, French Frigate Shoals, in 1991. Much to her amazement, she found that some female frigatebirds with young specialized in catching Sooty Tern chicks. The frigatebirds swooped down and snatched an unguarded tern chick to feed their own offspring. Shaped like miniature frigatebirds themselves, Sooty Terns would sweep into the air

Great Frigatebird juveniles at play. The "robber baron" of the bird world, the frigatebird chases other birds, forcing them to drop their prey, and then catches the meal in midair.

and cry frantically as a frigatebird carried off another chick. Other frigatebirds soon spied the action and jockeyed in the air for a chance to play steal the bacon.

Frigatebirds are designed for maneuverability and long-distance flight. They have the highest ratio of wing surface to body weight of any bird. In other words, they are mostly wing and tail feathers—aerodynamically designed to float like a kite, maneuver like a swallow, and dive like a falcon (I once saw one do a complete 360-degree somersault in the air as it dove). One cost for this mastery of flight is a reduction in their mobility on land. They cannot land on the water because they lack an oil gland to waterproof their feathers, so they must return to land to sleep roosting on shrubs. Their relatively tiny feet are useless for walking or swimming. Like bats, their weak feet can only cling to perches. The vulturine-looking birds nest in loose colonies where guano towers build up under their perennial perches, so the colony usually has a lot of flies in attendance.

Gray-backed Terns have the most eclectic diet of Hawaiian seabirds.[13] Besides plucking moths from flowers and eating lizards they catch

Great Frigatebird female with young on nest at Tern Island.

on the ground, they also eat Five-horned Cowfish.[14] Forty percent of their diet is composed of juvenile cowfish! Adults of these spiny reef fish seem unavailable to terns in the complex reef system, but it appears that cowfish spend their early life in the pastures of plankton at the sea surface near areas of upwelling currents, such as shoals and banks, and so become prey for these terns. Gray-backed Terns are pale versions of Sooty Terns, but instead of nesting in massive flocks, they nest in small scattered colonies. Lisianski Island has the largest colony in Hawai'i—up to 40,000 Gray-backed Terns.[15]

How much squid could a Sooty Tern suck, if a Sooty Tern could suck squid? In his book, *Seabirds of Hawaii*, Craig S. Harrison estimated that 10 million Hawaiian seabirds eat about 300 to 400 thousand metric tons of seafood each year. Squid accounts for at least half of the food eaten, fish about a quarter, and crustaceans the remaining quarter. Harrison concluded that seabirds eat over thirty times the amount of seafood taken by commercial fishermen and confirmed that the diets of boobies and frigatebirds are similar to that of tuna and thus of interest to biologists and fishermen alike.[16] Fishermen are already using information provided by boobies. When flocks of seabirds congregate at sea, the water beneath them usually holds tuna. Tuna drive the flying fish and squid to the surface, and fishermen, sighting birds feeding from far away, take their boats to the feeding flocks. Without following boobies and terns, fishing at sea is a hit-or-miss affair.

Determining fish consumption by seabirds requires an accurate assessment of bird populations, which is difficult to determine. Counting the numbers in large flocks requires practice and judgment skills acquired only over many field seasons. Unfortunately, this information is difficult to interpret because of the inadequacy of techniques descriptions, the lack of standardized and objective census techniques, and the inconsistency in

Sooty Tern feeding squid to its chick at Tern Island.

the duration and timing of most visits.[17] Random transects through colonies are the best way to estimate numbers. Using a line transect to count the numbers of nesting birds goes a long way to eliminating bias arising from picking areas of high or low bird densities. Data from transects are treated mathematically to yield population numbers. But the act of censusing itself can be harmful to the birds being counted. Accurately counting nocturnal burrowing seabirds is a near impossibility, so their total numbers will never be known. Some areas of islands are a maze of burrows, pits, and hidden honeycombs. Cave-ins can happen naturally, but human passage through colonies is always a disaster for the burrow-nesting petrels and shearwaters. Breaking through the ceiling of a burrow is a risk for bird and biologist alike. To save the bird from suffocation, the burrow must be reopened, but often clearing out one burrow collapses another, causing an unknown yet substantial amount of mortality. Properly conducted, repeatable surveys are a crucial research tool for the management of the marine resources in the Northwestern Hawaiian Islands. Seabirds can be used as an index to monitor the health of the entire ecosystem, because their populations are

known to fluctuate in direct response to fluctuations in the marine food chain.

The USFWS and NMFS are grappling with the best ways to manage fishing in the northwestern waters without adversely affecting the fragile fishing economy and marine ecosystems. But the depletion of fishery stocks in international waters will surely affect both the seabird and fish resources of Hawai'i. Fishing activities make some food resources available to seabirds that would not otherwise be. For example, the Black-footed Albatross is more commonly seen at sea than the Laysan Albatross, because it is an inveterate ship follower, but both species are alert for fish brought to the surface by fishing ships. However, scavenging behavior has its own risks. When foraging near fishing vessels, albatrosses can become tangled in transparent fishing nets and drown. The high-seas drift-net fishery, with 30-mile-long nets set in the North Pacific for Flying Squid, incidentally caught and killed about 100,000 Laysan Albatrosses and as many as 2 percent of the Black-footed Albatross population annually in the 1980s. In 1990, the United Nations passed a resolution calling for a moratorium on all large-scale drift-net fishing beginning in 1992, which temporarily eliminated the "bycatch" problem.

Following the drift-net ban was the emergence of the long-line fishery for billfish and tuna. The international fishing fleet has undergone tremendous growth in the early 1990s as countries sought new economic opportunities. Longliners maintain a "set" 60 to 80 miles long, with up to 3,000 baited hooks, deployed for 5 hours. Seabirds gather around the lines for easy pickings as the baited hooks are deployed or retrieved. Unfortunately, the birds get hooked and drown when they seize the bait or get tangled in the incoming lines. The fishery based in Hawai'i alone, ignoring the Alaskan, Canadian, and international components, killed an estimated 322,500 Laysan and 23,382 Black-footed Albatrosses between 1990 and

Black-footed Albatrosses fight over fish scraps in the North Pacific. Between 1990 and 1994, over 23,000 black-foots perished on longlines set in the North Pacific swordfish fishery.

1994.[18] The loss of Black-footed Albatrosses is of great concern because they seek out baited hooks and the Hawaiian population, with 41,090 breeding pairs, is down 15 percent in 8 years. Laysan Albatrosses are not as strongly attracted to ships as black-foots are, but they are also in a steep decline with 44,862 pairs.[19] Nevertheless, when a breeding adult is killed, the chick starves to death and the surviving parent takes several breeding seasons to find a new partner. Also, because more juveniles than adults are killed, future breeding stock is diminished.

An alarm has gone out regarding the endangered Short-tailed Albatross, because in 1995 at least two deaths were confirmed and one mortal-ity suspected due to fishing hook sets. Extrapolating to the entire industry, perhaps 1 percent of the world's population of this exceedingly rare bird, estimated to number around 1,200, is taken each year. Instead of examining and correcting the situation, the National Atmospheric and Oceanic Administration (NOAA) responded by requesting that the limit for taking short-tails be raised from one to four per year. NOAA did enact a ban on longline fishing activities within 50 nautical miles of land because endangered monk seals were getting caught and perishing in the gear. What is really needed is not a "Band-Aid" approach to management but a modified fishing gear system that delivers the hooks into the water rapidly so that albatrosses and seals don't have enough time to snatch at bait as the hooks are deployed.

Also needed is a clear boundary for the refuge, somewhere between the dotted line over 100 miles offshore that Theodore Roosevelt drew on the map in 1909 and a perimeter wide and deep enough to protect the coral reef ecosystem. Now is the time to clarify this. In May 2000, President Bill Clinton issued an executive order to establish a national system of marine protected areas. As if to underscore the need for such protection, a long-liner ran aground at Pearl and Hermes Reef the same month. The disruption of marine food webs from fishing and pollution needs to be mitigated now, before economic and political pressures become too great to permit effective action. And here is the crux of the problem. So long as the environment is viewed a "resource," it will be exploited until it is spent, or until a natural event like ocean warming renders it unmanageable. This is a lesson we ought to have learned in the past but seem prone to forget.

References

1. Polovina and Tagami (1980).
2. Boggs and Ito (1993).
3. He and Boggs (1995).
4. National Marine Fisheries Service biologist C. Boggs, 1997, personal communication.
5. Honda (1980).
6. Gary L. (Skip) Naftel, 1999, personal communication.
7. Gary L. (Skip) Naftel, 1999, personal communication.
8. Polovina (1994).
9. Polovina et al. (1994).
10. Polovina et al. (1994).
11. Ware (1995).
12. Seki and Harrison (1989).
13. Clapp (1976).
14. Harrison et al. (1983).
15. Harrison (1990).
16. Harrison (1990).
17. Knudtson (1980).
18. Kalmer and Fujita (1996).
19. U.S. Fish and Wildlife Service biologist Beth Flint, 2000, personal communication.

12

Gardner Pinnacles

He pūko'a kū no ka moana.

"A large rock standing in the sea."
Said of a person who is unchangeable and very determined.

Photo of Gardner Pinnacles taken on 21 April 1933.
(U.S. Navy, National Archives, Pacific Sierra Region)

Gardner Pinnacles is the least-visited island in the Leewards. Lying 117 miles west of French Frigate Shoals, Gardner Pinnacles consists of two rocks, the tallest standing 170 feet high and 200 yards long. A shallow underwater bank extends 10 miles to the southwest and about 5 miles in all other directions. This last bastion of basalt, approximately 3 acres in extent, was once about 80 square miles in size and may have been an island intermediate in size between Lāna'i and Kaho'olawe. Gardner Pinnacles is composed of fine-grained black basalt with calcite and olivine crystals as well as granules of augite and magnetite, all covered with a thick layer of odoriferous bird lime.[1]

The Pinnacles were first reported by Captain Joseph Allen of the Nantucket whaler *Maro* on 2 June 1820: "a new island or rock not laid down on any of our charts. . . . It has two detached humps. . . . We call it Gardner's Island."[2] Over the years, few explorers attempted to land on the treacherous rocks. Other vessels passing by gave it such names as Man-of-War Rock, Ballard Island, and Pollard Island. When the island was visited in 1826 by Lt. Paulding of the U.S. Schooner *Dolphin,* he wrote: "At three, P.M., on the fourth of January, a rock was reported from the mast head, eight leagues from us. It proved to be Ballard's Island, as it is called. At eight on the following morning, we passed within two hundred yards of it. It is about two or three hundred yards in circumference, and rises two hundred feet from the sea. On one side it has a considerable inclination, where seals had crawled up, and several were basking in the sun, almost to the very top. Large flocks of birds were perched on its ragged sides, or wending their flight around it.

Not the least sign of vegetation was any where to be seen."[3]

About a hundred years later, it was the Tanager Expedition that made the first biological survey of Gardner Pinnacles. A crew of eight landed and made their way to the summit with little difficulty. On the climb up, they discovered spiders, mites, moths, centipedes, isopods, and earwigs under the rocks and found only one plant species—the succulent sea purslane. Nineteen species of seabirds have been recorded at Gardner, twelve of which breed there. The most interesting species is the Blue-gray Noddy, which has so few breeding sites in Hawai'i.

Beginning in 1963, teams from the Pacific Ocean Biological Survey Program made four visits. They recorded ticks, flies, and beetles on the rocks and made more comprehensive bird surveys. Also in the early 1960s, the U.S. military made an unauthorized landing on Gardner Pinnacles as part of the HIRAN Project, one purpose of which was to determine the exact location of several of the Northwestern Hawaiian Islands for navigational purposes. Among other activities, the landing party blew off the top of the island to create a flat area for helicopter landings.[4]

Today, fishing vessels or research ships occasionally pass close to Gardner Pinnacles, but most ships steer clear for there is no safe anchorage and little leeway. The nearest anchorage is at Laysan Island, just over 200 miles to the northwest, but between the two islands lie several banks and reefs that can "eat" ships.

About a hundred miles to the northwest of Gardner Pinnacles is Dowsett's Reef. Captain Wood of the *Kamehameha V* thought he was safely passing through the region when he struck it hard at 3 A.M. on 11 July 1872. The reef was "not laid down in any chart in my possession . . . and is probably the same that the *Two Brothers* was lost on, over fifty years ago. I shall take the liberty of naming it Dowsett's Reef, after the owner of this brig."

(Captain Samuel James Dowsett had settled in Hawai'i in 1828. His eldest son, James Isaac, was the first Western child born in Hawai'i other than of missionary parents.) The story of the shipwreck was described in *The Friend* (1872), "a monthly journal devoted to Temperance, Seaman, Marine and General Intelligence" that carried Captain Wood's version of the accident:

[It was] A very dangerous place, and only one of many in a W.N.W. direction from French Frigate Shoals, to the coast of Japan. I was steering W by N at the time, fancying myself secure of Maro's Reef, twenty miles to the south. First saw breakers to leeward. Hauled to the wind, with courses up for an emergency. In twenty minutes she touched. Put the helm down and let go all halyards and anchored. Gave her 16 fathoms of chain when she swung to her anchor and fetched up on the rocks. Dark as pitch. Furled all sail and got a spare anchor and hawser ready. At daylight took it sixty fathoms ahead and hove her afloat. From the masthead, nothin [sic] in sight but sunken rocks in all directions. There was seven feet of water thirty feet from her stern, rocks close under the bow, and the question was, how in the world did she get there, and how was she to get out again? As the sun rose, it commenced to blow strong from the eastward, and at 1 P.M. it would be high water. Got a spring from the larboard quater [sic] and clinched it on the hawser as far ahead as possible; put a purchase on the spring and hove it taught. Loosed the jibs and courses; canted her head into a hole of deep water to the south with a ledge of rocks in all around; cut all clear and came out of the scrape. Now to get out of the pen. Went to the masthead and saw a hole in the reef, about 20 fathoms wide, that looked deep. Went though all right, with the lead going—ten fathoms—no bottom![5]

North of Dowsett lies Maro Reef, a gigantic underwater island that is awash at low tide. A

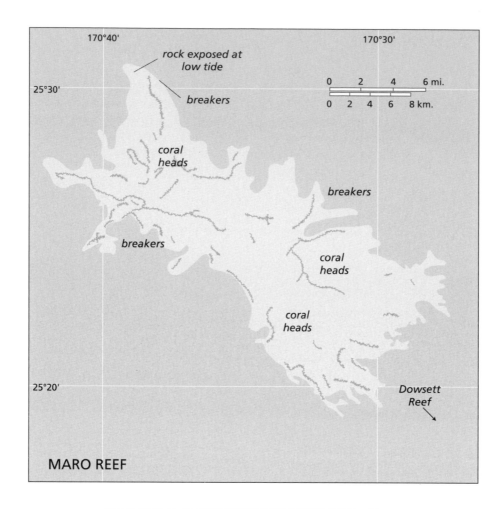

Chart of Maro Reef showing the size and shape of the former island.

small coral rock constitutes dry land on an oval bank 31 miles long by 18 miles wide. It was discovered in June 1820 by Captain Allen of the Nantucket whaling ship *Maro* after he discovered Gardner Pinnacles.[6] (Incidentally, *Maro* was the first of many whaling ships to enter Honolulu Harbor and one of the cofounders of the famous Japanese whaling grounds.)[7]

Unfortunately, I had only a passing acquaintance with Maro Reef. During a storm, the ship I was on, *Easy Rider,* sought shelter in its lee, 166 square miles of some of the most fertile marine areas in all of the Hawaiian Islands. The wealth of sea life at Maro Reef is matched only by that at French Frigate Shoals, and Maro Reef's proximity to Laysan Island, the most productive terrestrial ecosystem in the Leeward Islands, no doubt accounts for some of Laysan's fecundity.

References

1. E. H. Bryan Jr. (1942).
2. Clapp (1972).
3. Paulding in Clapp (1972:2–3).
4. King (1973).
5. *The Friend* (1872:31).
6. E. H. Bryan Jr. (1942).
7. Harrison (1990).

13

Laysan Island

'Ūlili alualu hu'a kai.

"Wandering tattler that chases after sea foam."
Said of a person who runs here and there for trivial things.

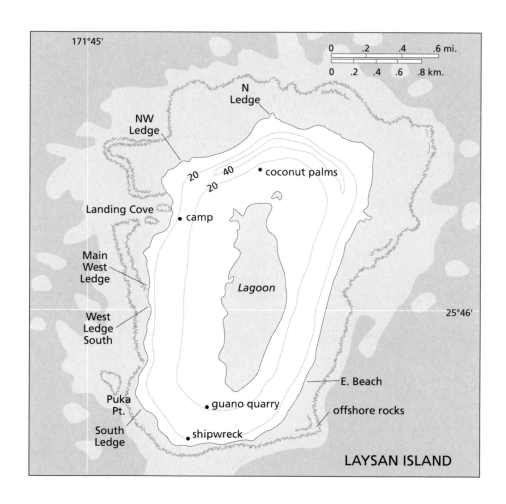

Map of Laysan Island.

Lying like a pearl in a turquoise setting, Laysan Island is the "gem of the Leewards." Lovely Laysan offers a glimpse of the paradise that was Hawai'i before humans arrived. Even after the destruction Laysan Island sustained at the beginning of the twentieth century, it still has the most impressive display of wildlife in Hawai'i. Laysan is an oceanic oasis for sea life as well as terrestrial land birds, plants, and insects. It is a refuge where reef sharks can swim slow figure eights in a courtship ritual tens of millions of years old; a sanctuary where monk seals can loll on beaches made of shell and coral, pulverized into powdery sand by endless breakers rolling in from across the electric blue Pacific; an asylum for unique species of birds that frequent the only lake in the Leewards. Albatrosses from the north, shearwaters from the south, sea turtles from the east, and biologists of each generation are all drawn to Laysan.

In 1979, I was making my first pilgrimage to Laysan. Black-footed Albatrosses circled our ship, the fishing vessel *Easy Rider,* as we plowed into the waves off Maro Reef. Our escorts personified the crew on board—rowdy and clamorous on land but subdued and silent at sea. We were only a day away from Laysan when *Easy Rider* (a ludicrous misnomer) ran into a front of heavy weather. As the albatrosses soared on 35-knot winds that ripped the tops off waves, the pounding boat coupled with diesel fumes drove me to the rail and then to the bunk for some rack wrestling. With every wrenching groan from the ship, I could visualize the aluminum plates popping off the hull. It was too rough to sleep, read, write, talk, think, or eat. With tempers frayed from ever-changing plans, there was little our team of biologists could do until the weather permitted us to travel through the reefs of Laysan.

As the low front passed, the waves smoothed out and we proceeded slowly through the coral heads. A passage through the maze on the northwest permits ships to anchor near shore in the island's lee. But if waves roll in from the west, then one can make a desperate dash through an obscure reef channel on the northeast side. Waves pushed by the prevailing currents curl in the channel before breaking short on the inshore reef. Luckily, we were able to land at the protected cove, which looks like a liquid Persian carpet.

All of our bunk rest came into play. We hustled to off-load about 7 tons of gear in as short a time possible so the *Easy Rider* could get out to the open sea. We hauled jerry cans filled with enough water for five people for a 3-month stint on Laysan, plus innumerable plastic 5-gallon buckets filled with food rations, personal gear, radio equipment—everything necessary to study birds and seals. We ferried all the gear from the surf up the sandy beach to the high and dry campsite before we permitted ourselves to be distracted by the splendors of Laysan Island.

A visit to Laysan Island is worth a week of seasickness. Make that 2 weeks I reckoned as I finally walked on the infamous isle. A ghost crab, fleeing my shadow, scurried down the beach and popped into its sand castle. I moved off the beach

NOAA ship Townsend Cromwell's *small boat at Laysan Cove. Seal biologist Brenda Becker has spent more time on Laysan than anyone else in recent years.*

to avoid a sleeping sea turtle. Lying by its side, a juvenile monk seal sneezed sand and flies from its runny nose. Venturing into the vegetation, I flushed the Sooty Terns whose cacophony of cries repelled me like a thief from a car alarm. The vast sand dunes of the northern shore were peppered with nesting Black-footed Albatrosses—the largest colony of this species in the world. The island interior was snow white with 280,000 Laysan Albatrosses, many nesting within a few feet of each other. The air was filled with flying objects, miraculously avoiding midair collisions. Even the ground was honeycombed with myriad burrows of the petrels and shearwaters.

Laysan hosts about 2 million breeding seabirds of seventeen species, plus two endemic species of birds, 240 monk seals, and numerous sea turtles and sharks. We had already been welcomed by the Laysan Finches. The inquisitive creatures hopped about the piles of gear looking for food. Indeed, camp needed to be finch-proofed for our mutual good. Finches would enter though holes in the tent netting, then fail to find their way out, so we had to shoo them out repeatedly. Individuals of this endangered, but locally abundant, species (the 1996 census counted over 14,000 on Laysan) were victims of their own curiosity, drowning in water cans or getting crushed by falling equipment. Our presence also had a positive effect as opportunistic finches followed us, ready to pounce on any unprotected seabird eggs. With a quick jab from their heavy beak, the yellow and gray birds puncture eggs and sip the runny yolk.

Laysan Island is roughly 200 nautical miles from Gardner Pinnacles and 900 miles from Honolulu. The second greatest single landmass in the Northwestern Hawaiian Islands, Laysan is a classic *motu,* or island of an atoll, elevated about 40 feet above sea level. Geologic uplift and coral growth have created an island that is shaped like a *poi* board. The island is approximately 1 mile wide and one and a half miles long—913 acres or 1.43 square miles in size. About 735 acres of fringing reefs grow from Laysan's shallow coral reefs, but the 100-fathom line encloses about 210 square miles of the underwater base.[1]

In the center of Laysan is a brown, hypersaline lake, about 100 acres in size. In 1859, it was up to 5 fathoms deep, but by 1923, it averaged 3 to 5 feet deep, with a maximum depth of 15 feet due to the accumulation of sand. The windward side of the island has major "blowouts" in winter when wind rips away the vegetation and sweeps sand into the acrid lake. On especially windy days, the water is whipped into a froth and billows of suds cover nesting albatrosses. The constant bubble bath causes albatrosses to abandon their cold, wet eggs. The lake is perched on a layer of nonporous coral rock thickly overlying the ancient basaltic volcano. Fresh water floats on the salt water and seeps to the surface around the lake and along areas of the coast where porous sand and impervious sandstone meet. The presence of Laysan Finches and Laysan Ducks drinking in certain areas is the best clue to finding fresh water on Laysan. A wide variety of migratory ducks, shorebirds, and gulls stops to feed and rest on the only lake in the Leeward Islands.

Laysan Finch sipping egg yolk.

Laysan Albatross eggs collected for harvest in the early 1890s. (Hawaii State Archives)

palm-trees on the island, and I collected twenty-five varieties of plants, some of them splendid flowering shrubs, very fragrant, resembling plants I have seen in gardens in Honolulu. I saw on the beach trunks of immense trees. The island contains about fifty acres of good soil. It is covered with a variety of land and sea birds; some of the land varieties are small and of beautiful plumage. Bird's eggs were abundant . . . there is a very small deposit of guano on this island, but not of sufficient quantity to warrant any attempts to get it."[3]

Unfortunately for the island's unique flora and fauna, Brooks was wrong about guano. Subsequent explorers did not fail to make the connection that abundant birds meant beaucoup bird droppings, aka guano. Chemically speaking, guano is phosphatized carbonate of lime. It is produced where birds digest fish bones and excrete uric acid. Ammonia is converted into nitrites by nitrification bacteria in the presence of lime in calcium nitrates. Bird excrement in shell and coral sand becomes phosphate rock after rainfall leaches most of the nitrogen from it. The subsequent phosphatization yields a high-grade guano —up to 80% phosphoric lime.[4] The most completely phosphatized guano is brown, dusty, and odorless and composes a large portion of the soil on all seabird islands, but a high-quality product occurs on only a few islands, one of which was Laysan.

In the late nineteenth century, British guano miners, following in the wake of the whalers, scoured the Pacific for seabird islands with workable guano deposits to supply the burgeoning market for agricultural fertilizers.[5] An Englishman,

With natural abundance and diversity, it was inevitable that Laysan Island would develop a notorious human history. Laysan Island was discovered by New England whalers, possibly Captain Briggs of New Bedford, cruising the Pacific in the early 1820s. The formal discovery is credited to the Russian ship *Moller* on 12 March 1828. Laysan was annexed to the Hawaiian Kingdom on 1 May 1857 by Captain John Paty of *Manuokawai* on his famous 50-day voyage through the Northwestern Hawaiian Islands. The ship's log described a pristine Laysan: "The island is 'literally covered' with birds. . . . Seal and turtle were numerous on the beach, and might be easily taken. They were evidently unaccustomed to the sight of man, as they would hardly move at our approach, and the birds were so tame and plentiful, that it was difficult to walk about the island without stepping upon them. The gulls lay enormous eggs. . . ."[2]

Captain N. C. Brooks of the *Gambia* described the island 2 years later: "There are five

Captain George D. Freeth and family, taken in the 1890s on Laysan Island. (Hawaii State Archives)

Captain George D. Freeth, convinced the German sugar plantation developer, H. Hackfeld, to finance guano mining on Laysan in 1890. The Hawaiian Kingdom granted entrepreneurs Freeth and Captain Charles N. Spencer the right to mine the deposits for a period of 20 years if they paid a royalty of 50 cents per ton. They were also granted the mining rights to other islands in the leeward chain, but besides Laysan, no significant amounts of guano were mined from them. "Under his [Freeth's] management many thousands of tons of guano were shipped to Honolulu and made into fertilizer, for the company had erected in 1894 a factory at Kalihi, Honolulu, covering several acres, and there, under the management of Dr. Averdam, a specialist, the guano was chemically treated and a fertilizer produced that answered the demands of the territory. . . ."[6]

Science followed commercialism. Henry Palmer and his assistant George C. Munro, bird collectors for the Rothschild Expedition, stayed with Captain Freeth on Laysan for 10 days in 1891. The purpose of that first expedition was to study the birds of the Northwestern Hawaiian Islands. The Honorable Walter Rothschild reported on their fieldwork in *The Avifauna of Laysan and the Neighbouring Islands,* a limited-edition folio published in 1893. The expedition was the first to scientifically describe three of the five endemic species of land birds from Laysan.[7]

In 1896, a gifted young zoologist and museum curator from Bremen, Germany, was granted a year's leave to collect specimens in the Pacific. Hugo Schauinsland began his trip in Honolulu and, with the help of H. Hackfeld and Company, gained an invitation to visit Laysan, perhaps his most important stop on an around-the-world collecting trip. The scientific name of the Hawaiian Monk Seal, *Monachus schauinslandi,* commemorates his trip and the rare seal he encountered. Dr. Schauinsland's account of his 3-month expedition to Laysan Island provides an important biological perspective on Laysan Island at a time before the ecosystem of the island was destroyed by guano miners, feather hunters, and rabbit depredations. His observations are unique; for example, contemporary biologists never knew that "The nicest plant on Laysan is, without comparison, a man-sized shrub, *Capparis sandwichiana.* . . . The splendor of the flower lasts only one night; it opens first at sundown and its life comes to an end the next morning when the sun has hardly raised itself above the horizon. . . . Its charm is also increased by its intoxicating highly pleasant fragrance. . . . Not too long ago, palms have also lived on the island, and, as the many remnants of their rotten stumps show, they were very numerous. However, the last living examples died off a few years ago. . . . It is not unlikely that castaways contributed to their demise."[8]

Guano diggers on Laysan Island, 1890s. (Hawaii State Archives)

Schauinsland reported that the avarice for guano was "the only reason why mankind visits this godforsaken island where only birds dwell." H. Hackfeld and Company, which leased the guano works at Laysan, provided the ship he sailed on. "The first reception on board was unforgettable!" he wrote. "The sailor's band, consisting of a kettle-drum, triangle and accordion, tactfully greeted us with the masterly performed, pretty song 'O'Susanna,' which is also very much appreciated amongst us on the banks of the Wiser."[9]

Schauinsland's observations from Laysan included a mix of biology, geography, and psychology. His style is quite lyrical, even flowery—atypical prose for a German scientist-adventurer:

As if by some stroke of magic, the whole appearance of the island was transformed. . . . One is awestruck by the bird's [Bonin Petrel] ability to accurately schedule its arrival time, almost to within an hour; where is the compass that guides its flight through the storms and hurricanes over the vast ocean toward this tiny speck of

land? . . . The animals on Laysan behave as they really are, without any fear. They had not yet learned to consider us their enemy, and therefore, we were constantly in a position (thus unbiased) to study not only their objective behaviour, but also and more specifically, their *emotional life* and their *spiritual character* . . . the chicks of the stoic albatross exhibit a quiet, agreeable, but somewhat feeble-minded disposition; quite the opposite from the pedantic, eternally fidgeting, sanguinistic tern. The black shearwater [Christmas Shearwater] however, has a decidedly melancholy disposition. . . . I can only compare to the sighing moans and cries of lament made by a very unfortunate person who is totally depressed about his life and the world.[10]

Perhaps Hugo Schauinsland was subconsciously referring to his wife, who accompanied him and did his dirty work. While Hugo was out collecting specimens, "day in and day out, she [his wife] would painstakingly prepare the specimens truly by the sweat of her brow as long as the daylight lasted. She suffered silently as cockroaches ate her fingernails while she slept and the tiny ants and beetles ate her labors of love, the specimens, and almost drove her mad. 'I often found my wife in tears over the lost efforts,'" he wrote. And all this after she had suffered a serious disease upon her arrival. For 8 days her condition was guarded, but she recovered only to immerse herself in preparing specimens with mercury, arsenic, and "old lace."[11]

Life was hell in the guano fields as well. Hawaiian and Japanese contract laborers worked

The guano mines, Laysan Island. (Hawaii State Archives)

at the end of a shovelful of bird guano, covered with dust and flies, in all weather. They cracked the guano rock with picks, crowbars, and sledges; shoveled it into wheelbarrows; and then rolled barrow loads to railway tramcars. Two mules pulled the cars down narrow-gauge railways to the storage sheds. About 100 tons per day were chuted through a hole in the wharf into a barge's hold, which then carried the material to the sailing ships anchored just offshore. Ships came two or three times a year. Mining was curtailed during the stormy winter months, when a caretaker watched over the island. In the spring of 1904, when the labor crews returned to dig, they found

that the caretaker had died while seated at his books.

It was not the first death on Laysan. On 11 August 1900, a labor dispute also ended in murder when the Japanese diggers refused to work any further. Only one man spoke both English and Japanese, and he told the group that they would be returned to Japan if they would not work. The laborers then pressed for more food and wages but were again refused. Brandishing tools and crude weapons, they took their case to Captain Spencer, the island manager. Four white men met the mutineers with loaded revolvers. Spencer told the crowd to move back, but his commands ap-

"In the last days of October, the first outposts of the magnificent albatrosses appear, and a few days later, from a higher vantage point, the island looks as if it were covered by large snowflakes. There is hardly a spot from which the dazzling white plumage of an albatross does not reflect back" (Udvardy 1996:21).

parently went unheeded. Several workers approached the overseers and someone panicked. When the smoke cleared, one miner was dead, one mortally wounded, and two more injured. The remaining thirty-eight were incarcerated to prevent further mayhem. During the 10-day court hearing, Spencer's assistant turned against him, calling him a drunkard and an ill-fit manager. He worked the testimony to his own advantage because he himself may have actually inflicted the mortal wounds. Character witnesses asserted that Spencer was not a drunk and, eventually, charges against him were

dropped.[12] Such were race and labor relations in 1900.

The U.S. Fish Commission Steamer *Albatross* visited Laysan in 1902. The expedition's Walter K. Fisher filed reports about the spectacular numbers of birds and first photographed the endemic species. In 1903, Professor William Alanson Bryan of the Bishop Museum visited Laysan Island and estimated that 10 million seabirds were present: "such an astonishing population, numbering perhaps twice the inhabitants of Greater New York."[13] But the guano digging had begun to affect the island

"Laysan is a true bird paradise; nowhere on earth is there another place like this. The land birds occupy an inferior position, enduring their role of barely being tolerated by the seabirds who are the dominant and ruling class here" (Udvardy 1996:19). Black-footed Albatross contemplates a Laysan Finch.

life. By 1911, just over a million seabirds were present. Several species of plants present in 1903 were extirpated from Laysan, including the last five *loulu* palms and the sandalwood that was once abundant on the northwest side. It was the beginning of the end for Laysan.

One man's name became synonymous with Laysan and linked with the destruction of island ecology. Ironically it was the man who loved the island the most: Maximilian Joseph August Schlemmer. He had arrived on Laysan as early as 1894 to be the foreman of the Japanese labor force.[14] One of his jobs was keeping the albatrosses off the tracks to prevent them from being crushed by

tramcars. In 1896, he was appointed superintendent of the guano operations. Luckily for him, Schlemmer departed Laysan before the Japanese laborers were murdered. He left to open a bar and boardinghouse on Kaua'i but returned in late 1900 to take Captain Spencer's place. In 1902, he gained notoriety for punching one Count Albert von Gravemeyer, who allegedly was trying to create more discord among the Japanese workers. The count sued Schlemmer for $3,000 in damages, but at the hearing the judge found in favor of Schlemmer, who, the judge decided, was just trying to keep the peace.[15]

Auburn-haired and fair-skinned, Max had arrived to live on Laysan with his beautiful wife Therese and three children from his first marriage, to Therese's sister, who had died. In 1900, Adam, his first-born son, died after just 12 days on Laysan. (A cross marks his grave near our camp. Beside him is the last resting place of a young Japanese woman who was the wife of one of the guano diggers. She died in childbirth, and her newborn baby was buried with her. The crosses, replaced in 1967, are conspicuous bird perches on an island with few trees.)[16] In 1903, Eric Laysan Schlemmer was born, followed in succession by five more Schlemmer children who were born on Laysan. Later, nine more children became part of Max's legacy of fecundity. Unfortunately, the same held true for his rabbits. Schlemmer released white domestic rabbits, Belgium and European Hares, and Guinea Pigs on Laysan, partly to amuse the children and as potential livestock for a meat-canning business—an action he would rue the rest of his days.

In 1904, the North Pacific Phosphate and Fertilizer Company, now known as the Pacific Guano and Fertilizer Company, sold their rights to Schlemmer for $1,750. The guano deposits were played out, after more than 450,000 tons had been removed. Substantial sums had been made selling the guano at $15 a ton. The successful speculative

Max Schlemmer and his wife Therese, with letter sent by W. A. Bryan of the Bishop Museum simply addressed to Max, c/o Laysan Island, Hawaii. (Courtesy of the Schlemmer family)

Crosses on Laysan mark the graves of Adam Schlemmer (d. 1900) and an unknown Japanese woman and child who died in labor in the early twentieth century. The crosses, where frigatebirds and noddies perch, were refurbished in 1967.

owned company changed its name to American Factors or Amfac, which became a giant Hawaiian corporation, one of the "Big Five," with vast land holdings in the Islands.) "The deposits on Laysan have been so depleted during the last few years that it does not pay the company to work the place. Early last year the island was leased to Max Schlemmer, for the past ten years 'Governor' of the island and manager there for the company. He is taking out, approximately 500 tons of guano a year, and this he ships to the Company."[17]

Known as the King of Laysan ("That sounds all right, but there is no money in it," Max was fond of saying), Schlemmer made a contract in December 1908 with Genkichi Yamanouchi of Tokyo. In return for $150 a month paid in gold, Max permitted them "to remove and sell freely phosphate, Guano and products of whatever nature from the islands of Laysan and Lisianski."[18] The Japanese were the main suppliers to the world's millinery trade, and they used this contract to export feathers, not guano. In 1909 they removed a ton of feathers and 2 tons of bird wings representing about 64,000 birds.[19]

Schlemmer's activities became illegal with the establishment of the Bird Reservation, and the U.S. Revenue Cutter *Thetis* was sent to investigate rumors of poaching. They found twenty-three Japanese on the island and guano sheds filled with sixty-five bales of bird wings, thirteen bales of bulk feathers, and thousands of rotting bird carcasses. The raid netted a ton of feathers and 119,000 bird wings estimated to be worth about $130,300. The Japanese were arrested and removed from Laysan, and Schlemmer, in Honolulu at the time,

venture had provided a major source of chemical nutrients for use in the Hawaiian sugarcane fields, which H. Hackfeld owned. (In 1915, to avoid jingoism at the onset of World War I, the German-

Laysan Albatross on tramcar tracks at Guano Operations Headquarters (date and photographer unknown). (Bishop Museum photo)

on Laysan. The shore party, headed by First Lt. Munter, noted the carnage that had occurred since their last visit. Blow flies and carrion beetles ruled the killing fields. "Between one hundred and fifty and two hundred thousand birds were found lying in heaps in all parts of the island. All of them were found on their backs with only the breast feathers missing. In the majority of cases the feathers had been pulled out, but in some instances knives had been used and the breast had been cut away from the bodies. . . . There are hundreds of eggs with young chicks in them that never hatched."[22]

Max Schlemmer could not stay away from Laysan, and though the government denied his appointment as a federal game warden, they agreed to his request to inhabit Laysan in the hope that his presence might prevent feather raids. In July 1915, he returned in the sloop *Helene* with his son Eric and a young sailor named Harold Brandt. They recolonized the island, repairing the buildings, and painting and repairing the cistern. He noted with regret the poor environmental conditions. "I think in time we will put the island in decent old condition if the government will listen to me," he lamented in his log, but ominously noted "the wind is blowing the sand away by the tons."[23] While laying in supplies for the winter, he took three seals and fifteen turtles for food and oil and pickled 350 albatross eggs with salt and lime. Yet Schlemmer and his companions ended up nearly starving to death.

While refurbishing camp, they noticed flotsam coming ashore from a shipwreck. The schooner *O. M. Kellogg*, en route to San Francisco, had run aground on Maro Reef, and the nine sur-

was indicted for poaching on a federal reservation and illegally importing contract laborers. When the two charges were dropped on technical grounds, Schlemmer retorted that the destruction of the birds was due to others and that if Territory of Hawaii Governor Frear had not prevented his return, he would have been there to stop the destruction.[20]

In his absence from Laysan, the rabbits he had previously harvested by hunting were uncontrolled. In just a few years, their appetites outstripped the vegetation growth, and the unique island ecology was forever altered. Rabbit depredations were so severe that the U.S. Biological Survey sent an expedition on the *Thetis* in the winter of 1912–1913 to eliminate them. The expedition ran out of ammunition after five thousand rabbits were killed, leaving several thousand more that continued to destroy the vegetation.[21]

In early 1915, the feather hunters returned. On 3 April 1915, the *Thetis,* with Commodore Salisbury at the helm, attempted to surprise poachers vivors made the 70 miles to Laysan in their life-boats. Max agreed to loan the captain his sloop *Helene* and they soon sailed to get aid at the nearest source—Midway Island—over 500 miles to the northwest. Max had provisioned *Helene* with supplies that he needed on Laysan and he had expected their prompt return. For weeks, Max and the boys waited in vain for their ship to return. Unfortunately, *Helene* had sunk in Midway Harbor during a storm and Max and the boys had only flour, sea turtles, fish, and bird eggs to eat.[24]

Log book entry, 15 November 1915: "Today we hoisted the stars and stripes for Mrs. Max Schlemmer's birthday. I sent Eric and Harold around the beach looking for some wreckage and to my surprise, they came home with a tin of dry potatoes which washed ashore on the east side of the island. It has some saltwater in it but was mostly dry in the middle."[25] The tin of potatoes floating in from the *Kellogg* wreck was a godsend and was the first potato they had tasted in 4 months. The three persevered, prayed for deliverance, and were rescued in the nick of time by the USS *Nereus*. Taken to Honolulu, Max ran into unexpected trouble. World War I had erupted and he was accused of being a German spy using Laysan as a listening post. The insult to his patriotism was the final straw. Max quit the sea and never saw Laysan again. He became head of maintenance for Amfac until his retirement. The "King of Laysan" died in 1935.

Today his legacy lives on. Max's youngest daughter, Helene Schlemmer Brown of Waimā-nalo on Oʻahu, recounted the family history for me in her home. "My dad's sloop, the *Helene*, sunk in 1915 off of Midway and since I was born in January 1916, after the sloop sunk, I was named after the sloop." At 85, Helene's spirited determination and *aloha* spoke of her inspiring father, a courageous man of the times, a resourceful capitalist in old Hawaiʻi, and I came to understand that old Schlemmer, the scoundrel who devastated Laysan,

was simply a man of his times.

References

1. Ely and Clapp (1973).
2. *The Polynesian* (1857:40).
3. Ely and Clapp (1973:21–22).
4. Skaggs (1994).
5. Skaggs (1994).
6. *Pacific Commercial Advertiser* (2 July 1906:12) in Apple (1973:86).
7. Bailey (1956).
8. Udvardy (1996:15–16).
9. Udvardy (1996:5).
10. Udvardy (1996:21–22).
11. Udvardy (1996:31).
12. Ely and Clapp (1973).
13. W. A. Bryan (1911:309).
14. Farrell (1928).
15. Schlemmer family tapes and letters, Waimānalo, Hawaiʻi.
16. Schlemmer family tapes and letters.
17. *Pacific Commercial Advertiser* (2 July 1906:12) in Apple (1973:87).
18. Ely and Clapp (1973:39).
19. Ely and Clapp (1973).
20. *Pacific Commercial Advertiser* (2 July 1910) in Clapp et al. (1996).
21. Ely and Clapp (1973).
22. Ely and Clapp (1973:45).
23. Schlemmer family tapes and records, Waimānalo, Hawaiʻi.
24. Schlemmer family records, Waimānalo, Hawaiʻi.
25. U.S. Fish and Wildlife Service tape recording of Eric Schlemmer reading Max Schlemmer's log book in interview with U.S. Fish and Wildlife Service, 1971(?).

14

Utter Desolation

Puaea ka manu o Ka'ula i ke kai.

"The birds of Ka'ula die at sea."
Do not wander too far from home lest you be destroyed.

Tava and Keale, *Ni'ihau: The Traditions of an Hawaiian Island*

When the USS *Thetis* again visited Laysan in 1916, First Lt. Munter noted that the rabbits had multiplied and were difficult to capture. He recommended to his superiors that they be removed before Laysan became "but a sand spit like Lisianski Island."[1] By 1918, the rabbits had almost eaten themselves out of existence—not more than a hundred remained. The rabbits had also eliminated twenty-six species of plants in 20 years.[2] It was also too late for the Laysan Millerbird, which became extinct around that time. Millions of years of evolution were vanquished in two decades.

The millerbird was once abundant and conspicuous, attracting much attention through its trusting ways. The birds perched on people and sang unconcerned, picked moths off the dinner table, and flitted around laboratory equipment. In June 1891, George C. Munro described how miller-

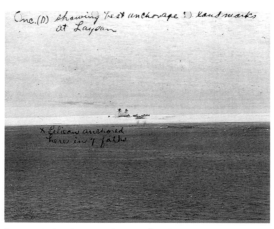

Laysan Island was a desert island in May 1924, when the USS Pelican *visited with Hawaii Bird Reservation warden Gerrit Wilder.* (National Archives, Pacific Sierra Region)

birds entered the guano miners' buildings in quest of moths attracted to the lantern lights. Giving chase to moths, the birds even broke test tubes in the guano laboratory but could slip away without being caught.[3] In the end, their innocence failed them. Tragically, the species was extinct by 1918, even before the vegetation was entirely consumed by rabbits.

It was only after the Laysan Millerbird was extinct that the Nihoa form was discovered. Nihoa Millerbirds are less trusting, perhaps from contact with prehistoric Polynesians, because they managed to stay hidden until 1923. Nihoa and Laysan Millerbirds are so closely related that some conservationists have suggested that some Nihoa Millerbirds be transplanted to Laysan as insurance in case of a fire or introduction of pests to Nihoa. Sheila Conant feels that transplanting this subspecies could be problematic because the Laysan habitat was bunchgrass and Nihoa's is shrubs. As discussed in chapter 2, Dr. Conant's other research with Nihoa and Laysan Finches demonstrates that genetic changes can develop quickly, and biologists contemplating transplants may be tinkering with evolution in the Northwestern Hawaiian Islands.[4]

By 1923, Laysan was a wasteland. The Tanager Expedition, a joint effort between the Bureau of Biological Survey (then of the U.S. Department of Agriculture), the Bishop Museum of Honolulu, and the U.S. Navy, brought a biological "swat" team to finally rid Laysan of the rabbits. Alexander Wetmore of the National Museum of Natural History in Washington led the Tanager Expedition. (Dr. Wetmore went on to become the director of the Smithsonian Institution and an internationally renowned ornithologist.) Wetmore was assisted by Eric Schlemmer, son of Max. Born on Laysan, Eric provided Wetmore with a unique perspective on Laysan. Eric also helped E. C. Reno and E. L. Caum kill over two hundred rabbits, replant vegeta-

tion, and release Laysan Rails brought in from Midway.[5]

Donald R. Dickey, the cinematographer of the Tanager Expedition, documented the dying ecosystem. His diary supplemented his film recording of the scene. "Verily the damned rabbits have done their worst," wrote Dickey. "As far as I can see with the glasses and from our hurried trip down the island, there is <u>not a living bush or twig</u> or <u>spear</u> of <u>grass left</u> on the whole island outside of the two poor coconut trees and 3 bushes near the house. . . . In my wildest pessimism I had not feared such utter extirpation of every living plant." Dickey was able to record the essence of Laysan evolution in its last moments: "Set up the camera and soon had a Laysan Finch recorded on film (if all goes well and this cursed <u>100%</u> humidity does not wreck me.) As I finished I heard a weak but charming song behind me and whirled to find one of our pair of Laysan Honey Eaters [Honeycreepers] singing his heart out for me. Whirled the camera, slammed the focus lever, cranked and think I have him. And before I had recorded the footage and "shot" Schlemmer came up holding a Laysan Rail in his hand. To get all the footage possible we turned him loose in front of the high speed. Meantime I had held alive and unhurt in my hand one of the two Laysan Rails we know are left on the island. . . . To think of getting one of the 3 honeyeaters we know to be alive and one of the 2 rails and one of the finches in lightning succession was indeed luck."[6]

Bad luck soon followed, brought on, perhaps, by earlier actions of the expedition members themselves. During a shark-killing spree at Kure Atoll on 16 April 1923, over thirty sharks had been

Desolation caused by rabbit depredations on Laysan Island. A few Wedge-tailed Shearwaters persist. April 1923. (Bishop Museum photo)

Laysan Millerbird at nest. Photo by Walter K. Fisher, 1902. (Denver Museum of Natural History Photo Archives.)

shot out of a school of at least five hundred. Later, on Laysan, as stiff trades blew from the northeast on 25 April, Wetmore recalled that "Old sailors say the wind is due to the many sharks killed at Ocean Island. The Hawaiian shark god is offended and shows his displeasure in this manner."[7] While the scientists were pinned down, a ferocious sandstorm blew the last three "Laysan red-birds" into oblivion. The Laysan Rails died off in the aftermath. The Laysan Ducks and Finches survived because the ducks ate brine flies at the lake and the finches scavenged on dead birds. As Dickey struggled under the inclement conditions for filming, he wrote of his frustrations and exaltations. His diary is worth quoting at length:

———————

Monday 4–23–23. Rain & mist—driving N. gale—clearing to sandstorm in afternoon. Hell of a day. And now I must pay for my fun! Awoke to the shattering tune of the tent straining at every peg.

... Cursed the Sahara of Laysan and decided to declare it Sunday. ... Saw 15 Laysan Teal all in sight at one time, and Ball afterward reported 18 when he passed them earlier—they are paired off now, but congregate at some small holes NE of the lagoon which seem to hold the rain water in less brackish form than elsewhere. The finches are common about these same holes. The males are wearing into the full yellow plumage and the boys have reported seeing them carrying nesting material into crannies in the guano rocks. ... This afternoon the clouds broke at times, but the gale increased until the whole island seemed to be walking—a regular cursed sandstorm of the worst sort. ... Off the south end of the island flaps a regular sand "banner." Life is Hell! ... to Hell with it and its islands, is the way I feel tonight. Tried scrambled Man-o-war bird egg tonight. It was doubtless nourishing, but I did not relish my share.

———————

Tuesday 4–24–23. Cloudy. North gale—sand storm. Hell of a day. The gale and blinding sandstorm continues unabated and life in the open or in the tents or tumble-down shacks is equally unbearable. Not much sleep for anyone, but luckily my tent has held so far, by turning the fly loose. ... Everyone except smiling George, the cook, is on edge. Took Eric and plowed out into the stinging blast of sand. ... There is not a shearwater burrow entrance visible on higher sand ridges and the island seems almost deserted of birds. A Man-o-war colony that had a dozen eggs yesterday and another dozen mated couples is deserted except for one male and one female that are sticking to their eggs. The rest have given it up and taken to the air leaving a feast for the curlew. . . . No sign of the teal today.

———————

Wednesday 4–25–23. Broken clouds—NNE gale—sandstorm (third day) hell on earth. The terrific cold gale and sandstorm continues unabated—almost unbearable—getting on our nerves—third day however, wind a trifle E of N and scudding clouds broken with moments of sun, so we pray to heaven it may go down with the sun tonight. Outside sand cuts face like a knife. . . . No sign of let-up tonight but sharp rain squalls are laying the sand.

———————

Thursday 4–26–23. N. gale unabated—rain squalls —less sand blowing—hell continued.

———————

Saturday 4–28–23. Heavy Rain. . . . At last the wind dropped after the worst wind siege I have ever experienced, but only to veer to the east and bring up the first prolonged torrential downpour we have had. . . . We could have caught a cistern full of water had we had a cistern instead of pans and pots and coal-oil cans. By 9:00 it had cleared gloriously with the second collection of decent clouds for photography we have had on the island.[8]

———————

The storm was over and the most grievous loss was that of the Laysan "red-bird," *Himatione sanguinea freethii*, named for the original colessor of Laysan. This brilliant honeycreeper was a subspecies of the crimson 'apapane, one of the most common native birds in the rain forests of Hawai'i today. The Laysan 'apapane, as it is now called, fed on flower nectar and insects, especially from the broad flats of the succulents bordering the lake. "Here they may be found throughout the day *walking* around after small insects or drinking honey from the blossoms. The brush-like tongue of the

himatione renders the gathering of honey an easy task. It is not uncommon to see one go from flower to flower and insert its bill between the petals of a nearly blown bud with a certain rapidity and precision which suggests a hummingbird, except of course the fact that the himatione is on its feet."[9]

Unlike the millerbirds, which ate moths whole, the honeycreepers fastidiously picked off the wings of moths, always holding them in their left foot, and ate only the soft bodies. Honeycreepers visited the shacks of the guano miners and drank from leaks in the rain barrels. They were not as common as the other Laysan land birds but neither were they rare. One scientist recorded its presence and how, when captured in the hand, it still sang fearlessly. "A most touching thing occurred; I caught a little red Honey-eater in the net, and when I took it out the little thing began to sing in my hand. I answered it with a whistle,

The last photo of a dying species—the Laysan Honeycreeper, taken by Donald R. Dickey, 1923. (Denver Museum of Natural History Photo Archives. All rights reserved)

which it returned and continued to do so for some minutes, not being the least frightened."[10]

That the honeycreeper existed on an atoll is testimony to the relative diversity and stability of this raised and expansive island environment, safe from storm surges. But how it arrived on Laysan in the first place merits some speculation. Initially, it was considered a relict species from when the island was a massive forest-clad volcano. This is now considered impossible because the first honeycreeper evolved about 3 million years ago and Laysan is millions of years older than that. Did an *'apapane* fly in from the rain forests of Kaua'i, 600 miles away? If so, why isn't there a subspecies on Nihoa, which is closer and has more vegetation? Did the Polynesians destroy its habitat there? Only questions remain, because the last three Laysan *'apapane* were blown off the face of the earth. Still, the living image of one survives. Dickey captured its swan song on film and preserved its inspiration. "The tiny honeycreeper was probably the most specialized in its feeding habits of all the endemic land birds on the island. God knows when the last flower bloomed on this barren waste, yet here are at least three individuals of this specialized form persisting as sort of heritage from the last nest of the species that was built in sufficient cover to survive. But as it seems to me, old age and death now inevitably stalk this childless remnant of a vanishing species.... They had done their best to adapt themselves to the changed environment and were scrabbling about among the rocks and on guano earth picking up small flies. Their charming song

is out of proportion to their size. Altogether they cheered my morning immensely."[11]

A creature as sensitive as the honeycreeper never had a chance compared with the durable Laysan Rail. Yet the saga of the Laysan Rail is perhaps a greater tragedy because it ultimately died of "red tape." The sandy brown, red-eyed rail was endemic to Laysan, where it was a conspicuous moth catcher. Rails boldly foraged around the houses at night chasing insects attracted to the lights. They also ate caterpillars, maggots, and seabird eggs, which they broke open themselves, and even sought out meat scraps from the biologists preparing bird specimens. Rails were opportunists,

Laysan Rail on the ground. Photo by Alfred M. Bailey, 1913. (Denver Museum of Natural History Photo Archives. All rights reserved)

114

and scientists observed them chasing Laysan Finches away from seabird eggs the finches had plundered. The rails took over the food sources with much bluster. They were abundant on Laysan, reportedly running over the ground like mice, and were remarkably tame. In spite of their active nature, the 6-inch-long "wingless birds" were not collected until 1891.[12]

Ship captains successfully transplanted rails to predator-free Midway Atoll in 1891. By 1905, rails were as common on Midway as they once were on Laysan, numbering between two thousand and five thousand. Their evening chorus from the cover of bunchgrass sounded like a handful of marbles bouncing down a glass roof.[13] Alfred Bailey and George Willett, members of the Bureau of Biological Survey, also took Laysan Rails to Lisianski Island in 1913. "In mid-February we began to capture rails to transplant on Lisianski, and to add fresh stock to those already on Midway. It proved easy, for we merely took a little box and a six inch stick to hold up one side of it. A chicken egg—which the rails could not break—was placed for bait, and when a half dozen birds were inside, jumping off the ground to give more force to the beaks' strike on the egg, we merely pulled the string. We caught more than one hundred, and liberated them on Lisianski and Eastern Island [Midway]. It is doubtful if they survived long on the former island, for the vegetation was disappearing due to the ravages of rabbits."[14]

Indeed, the Laysan Rails were losing their habitat to the ravages of the rabbits. In April 1923, Wetmore attempted to save the species by reintroducing eight rails to Laysan from Midway. The last of these rails lived on Laysan until mid-May, but because the destruction of the island ecosystem was complete, the species failed to survive in its home habitat. That the rail became extinct is especially tragic because it came so close to being reintroduced to Laysan again. Regular patrols after 1923 by the revenue cutter *Thetis* and revegetation

efforts helped the habitat to begin to recover. The vegetation appeared to be replaced on Laysan by 1936, albeit with a loss of species diversity. Bird populations, including the Laysan Finch and Laysan Duck, generally returned to preexploitation levels. Rails would have thrived if they had again been reintroduced. But their reintroduction was influenced by the nature of the finch.

With its outsized beak, the Laysan Finch is a survivor because it is an opportunist, not a specialist, like the millerbird or honeycreeper. When the vegetation vanished from Laysan, the finches survived on seabird eggs and rabbit carrion, though their "normal" diet includes seeds and succulent plant parts and whatever else they can find that is edible. "One of the last birds to disappear from the island will be the finch," observed a scientist in 1912. "Laysan Island is the ideal place for this bird, but should anyone be rash enough to introduce it to a civilized community it would be a pest that would rival the English Sparrow."[15]

George C. Munro, with the aid of Thomas M. Blackman, who was stationed on Midway in 1940, proposed that Laysan Rails and Laysan Finches be brought to Honolulu and kept in captivity as subjects for bacteriological tests. Munro had a theory regarding the role of disease in the

Laysan Finches drinking from a freshwater seep, suggesting their "pestiferous" nature.

Laysan Duck and ducklings feeding on shore flies at Laysan's lake.

disappearance of certain Hawaiian birds and wanted to test the species from an isolated island. (White Terns were also considered for release in windward O'ahu.) Thomas Blackman wrote to Edwin H. Bryan Jr., curator of collections at the Bishop Museum, on 4 April 1940: "I already have permission from Rear Admiral Murfin to take not more that 12 birds of each species," to which Bryan replied: "At some time in the past, acting on the recommendation, I believe, of the U.S. Biological Survey, USBS, (or perhaps of Dr. Alexander Wetmore, who was then connected with the USBS, and is now with the Smithsonian Institution), the Hawaii Board of Agriculture and Forestry passed

a ruling which barred the importation of the Laysan Finch. Importation of this bird could not be made without reversal of this ruling; Mr. E. L. Caum, is against such a reversal, and, from what I have heard . . . it would not be good to have the birds at large in the main islands."[16]

Bryan went on to explain that Ed Caum, warden of the Bird Reservation under the U.S. Biological Survey (as well as a former member of the Tanager Expedition), objected to the rail being brought to O'ahu. He believed the rails could not survive in captivity or at large because he had tried to maintain some that he brought back in 1923. They would not breed for him and were so temperamental that any upset was fatal. Why, he argued, should specimens be sacrificed in an attempt to repeat a failed experiment? Furthermore, Caum believed that taking specimens of this rare species would set a bad precedent, which might lead to wholesale requests to the Navy for specimens of the Laysan and Wake Island Rails.

Blackman countered and in subsequent correspondence with Bryan wrote: "it seems strange that anyone with scientific tendencies should argue that because an experiment has been tried and failed, it is useless to try again, in quite probably different conditions. . . . I wonder if it is generally known . . . how extremely abundant both the Rail and the "Finch" are over the whole area of both Sand and Eastern Islands, they cross one's path wherever one goes at all hours of the day and run under buildings in our camp. There must be literally millions of them . . . but when the term "rare species" is used it is very misleading to those unacquainted with its extreme abundance on these two islands, and would be hard to imagine what the harm the taking of not only twelve but a hundred or so could possibly do if they were to serve any purpose of research, or even zoological study. I still hope that some discretion may be shown before my leaving here, which will be around June 1 [1940]."[17]

Blackman mentioned that all the permits from the Navy were in order for the transfer of Laysan Rails, Laysan Finches, and White Terns, but in the end the U.S. Biological Survey refused all three on the basis of concern about the finch only. Blackman left Midway in June and the rails stayed behind, but the idea did not die with him. In late 1940–1941, Munro requested that Walter Donagho capture rails on Midway for reintroduction to Laysan Island. Donagho reported that he had found a nest with four eggs on 20 June 1941 and that adults were plentiful on Midway. "I had at least 20 rails in a cage ready for him [Munro] when I received a letter saying that Mr. Locey of the Board of Agri-

culture and Forestry, and the boss, had said 'no.' Subsequently, the rails were wiped out by rats. . . . I blame Mr. Locey directly, for the extinction of the Laysan Rail."[18]

The rest is history. The Battle of Midway occurred in 1942. By the next year, Black Rats and mosquitoes accidentally arrived with the military, and the last rail was seen on Eastern Island in June 1944, ironically by Ed Caum, the man who had first refused their export to Honolulu. The final resting places of Laysan's extinct birds are drawers in museums scattered around the world, where about 90 honeycreeper, 82 millerbird, and 140 rail specimens represent the lost Laysan populations.[19] The only place the public can view the lost magic of Laysan Island is in a diorama at the Denver Museum, which contains several mounted specimens of the extinct land birds.

Thankfully, there are two bright spots in the Laysan ecotragedy. There are still healthy populations of the Laysan Duck and Laysan Finch. Formerly referred to as a teal because it is smaller than most ducks, the Laysan Duck is a species derived from Mallard stock. Scientists have discovered that the Laysan Duck wing is malleable, like the beak of the Laysan Finch, because each wing has nine flight feathers while the Mallard has ten per wing. The species may be demonstrating one method by which flightlessness could develop in island birds. Ground-dwelling birds that live on predator-free islands do not need their wings for escape. They rely on their legs to get around, preferring to run away from what little danger there is and only flying as the last resort. Perhaps because of the fickle ocean winds, the ducks only fly short distances so that they don't end up lost at sea.

Ducks are curious and were easily hunted for food and feathers by early visitors. Thus, their populations had begun to plummet even before the rabbits arrived and continued to fall after the rabbits were gone. The duck population was once

Laysan Duck paddling across the lake at Laysan at sunset.

down to six adults before the sandstorm in 1923. Two years later, Wetmore saw twenty-five ducks and the population continued to vacillate.[20] By 1936, there were only eleven ducks present, but the species slowly recovered to several hundred. The population currently fluctuates between one hundred and six hundred individuals, evidently a result of inbreeding, severe winter storms, dry lake shores, and differing survey techniques. A hurricane hitting Laysan dead-on could conceivably blow all these ducks out to sea. However, captive populations thrive in aviaries worldwide and appear to do well despite a diet without brine

flies, so birds from these sources could be used to restock Laysan if it ever becomes necessary.

The lake at Laysan is critical habitat for the duck. With about 15 percent salinity, the lake holds several invertebrate species not found elsewhere in the Northwestern Hawaiian Islands. Underwater, the margin of the lake is red with brine shrimp that were probably first carried to the Laysan lake as dried eggs stuck on the legs and feathers of migrating waterfowl. Swarming on the shore are brine flies that lay their eggs in underwater algae. The submarine larvae develop into adults in a bubble until their wings are fully formed. They then pop up to the surface and fly off to join the phalanx of flies on the mud. The Laysan Duck has developed a unique behavior to harvest this mobile protein source. The duck tilts forward with its neck outstretched and runs along the lake shore. As the flies lift off the sand to avoid being stomped by the charging duck, the bird opens its mouth to snag flies in the air. After the bird passes, the flies resettle on the sand, only to be disturbed as the duck charges back again.

References

1. Ely and Clapp (1973:47).
2. Christophersen and Caum (1931).
3. Munro (1944).
4. Morin et al. (1997).
5. Olson (1996).
6. Olson (1996:124, 137).
7. Olson (1996:43).
8. Olson (1996:139–143).
9. W. K. Fisher (1903) in Bailey (1956:119).
10. Palmer (1891) in Bailey (1956:119).
11. Dickey (1923) in Olson (1996:129, 132).
12. Ely and Clapp (1973).
13. Bailey (1956).
14. Dill and Bryan (1912:89).
15. T. M. Blackman correspondence (1940) with E. H. Bryan Jr., Pacific Science Information Center, Grp 1, Box 20, Bishop Museum Archives, Honolulu.
16. T. M. Blackman correspondence with E. H. Bryan Jr.
17. T. M. Blackman correspondence with E. H. Bryan Jr.
18. W. Donagho, personal correspondence with Craig S. Harrison, 24 June 1983, Alexandria, Virginia.
19. Ely and Clapp (1973).
20. Munro (1944).

15

The Weed War and the Cold War

Noho ke koa'e i ka lua.

"The tropicbird remains in the hole."
Said of one who does not express his opinion.

Brown Noddies on nest in sandbur on Laysan Island.

Eric Laysan Schlemmer returned to his birthplace in 1971, escorted by Eugene Kridler, the first U.S. Fish and Wildlife Service (USFWS) refuge manager in Hawai'i. Schlemmer remarked how the vegetation had changed the original look of Laysan and he noted how many weed seeds now clung to the down of albatross chicks. Vegetation change is a dynamic process and goes almost unobserved, yet Schlemmer noticed that new plants were infiltrating the island.[1]

At least 312 plant species have been recorded from the Northwestern Hawaiian Islands. Most of them are alien species introduced to Midway; only 37 are indigenous, found throughout the tropical Pacific; and 12 are endemic species, found only in the Leewards.[2] More than one-third of the 28 plant species found on Laysan are introduced. A number of them pose serious problems, especially common sandbur, a grass from Central America that arrived in Hawai'i about 1867. Covered with hooked spines that cling tenaciously to clothing, skin, and feathers, sandbur is easily transported. Tern Island, Laysan, Lisianski, Kure, and Midway are now infested with sandbur. Nihoa was lucky. A few sandbur plants were found and removed before the species could spread.

Sandbur may have first arrived on Laysan in 1961, when the U.S. military established an operation to determine first-order astronomic stations and azimuth marks.[3] Air Force and Army personnel camped out for several weeks at a time, hauling in thirty fuel and water barrels. They inadvertently introduced the aggressive sandbur and Hairy Horseweed.[4] Later, the Coast Guard was requested by the Pentagon to place target tarps on Laysan in 1964, presumably to check high-altitude photography. The tarps were supposed to be sanitized but were not. Over 25 pounds of weed seeds were collected from them. The tarps were placed on Laysan and removed several weeks later by the Coast Guard. Refuge managers were successful in controlling sandbur until 1977 when Laysan became semipermanently inhabited by monk seal and seabird biologists. In spite of their efforts to limit the spread of alien species, it appears that biologists inadvertently carried the ubiquitous sandbur when making their rounds to their study sites. Sandbur may have also been assisted by Laysan Finches. They eat sandbur seeds and may get hooked by seeds that then hitch a ride to another location.

By the late 1980s, sandbur had expanded throughout Laysan and native bunchgrass declined. Old photos show extensive fields of bunchgrass, especially where we were camped, revealing the changes that were slowly occurring. Sandbur crowds out bunchgrass, which provides the main nesting areas for Laysan Ducks, Laysan Finches, and burrowing seabirds. Seabirds cannot dig burrows under sandbur as easily as they do under bunchgrass because the shallow roots of sandbur do not hold the sand as well.

In 1991, the USFWS declared war against sandbur. Almost 30 percent of the vegetated area on Laysan or about 10 percent of the entire island had become colonized by the weed. Almost 100 years after guano miners first arrived on Laysan with picks and shovels, a team of biologists began to use similar tools but for a different end. This was the first skirmish in the sandbur war, and I got an idea what the guano miners experienced in the hot interior of the island. With picks and "McClouds," a forest firefighter's friend (but an especially heavy and unwieldy tool), we dug up mature clumps of sandbur, then raked, dried, and burned them. The magnitude of the infestation and the amount of seed stock in the ground was overwhelming. We dug, burned, sprayed, and rooted out the weeds for weeks, barely making a

dent in the population. According to the farmer's adage: "one year of seeding, seven years of weeding." At our rate, it would take 210 years to finally rid Laysan of sandbur!

Weeding was brutal work. On glaring hot days, when the trade winds slackened, the July sun beat down. The heat rising off the sand was enough to dissipate rain-bearing clouds over the island. It got so hot that the addled eggs of albatross exploded spontaneously, splattering black goo and shell fragments like malodorous shrapnel. Hard work in the sun required plenty of fluids. It seemed impossible to drink enough, even though, fortunately, we had a solar still so water was not limited. It probably was for the guano diggers, who relied on a brackish well and rainwater runoff from their tin roofs.

Since my stint in 1991, teams of biologists have occupied Laysan continuously, slapping flies and fighting the weed war to the tune of almost 1 million dollars. In outlying areas of the main infestations, known affectionately as the "Blob" and the "Amoeba," teams pulled up seedlings and sprayed plants with herbicide on a rotating 2-week schedule.[5] By 1996, many areas had been initially cleared, but they required constant vigilance. With this level of effort, it appeared impossible to eradi-

Weeding and hacking away at the sandbur with "McClouds."

Beth Flint holding a Bonin Petrel rescued from a collapsed burrow.

efforts. The USFWS hopes to bring Nihoa Palms to Laysan and, eventually, Nihoa Millerbirds may follow, so that there is a hedge against extinction in case diseases, predators, fires, or hurricanes wipe out the Nihoa population. Perhaps the rail can be resurrected from the Baillon's Crake, a type of rail from the islands of southwestern Oceania.

During the weed war we received a crash course in terrestrial invertebrates. Hundreds of flies swarmed over our sweating bodies, and large spiders carrying egg cases ran over our feet. The introduced house flies are especially numerous on Laysan Island during the summer. The accumulated carcasses of dead albatross chicks and rotten eggs near the end of the breeding season provide abundant food. House flies crawled over everything, especially plaguing the monk seals as they sought moisture from the seals' nostrils, eyes, and open wounds. The seals fled into the water to be free of the flies and we escaped to fly-proof tents. The flies carry diseases such as avian pox, which infects many albatross chicks and adults, causing knobby growths about their face and beak. Many birds die during periodic epidemics of pox, especially on Midway, where dense, moist grasses provide optimum fly breeding habitat.

Louse flies, or hippoboscids, are part of the native insect community that we had to learn to live with when we worked near booby and frigatebird colonies. Rubbery and slow-moving, these flat-bodied flies slip under the feathers of frigatebirds and boobies to suck their blood. When a frigatebird flies off, it leaves a swarm of homeless flies in its wake that land on the next warm-blooded animal they find. Trying to remove hippoboscids from the inside of your shirt is merely annoying, however, because they don't bite people.

But ticks do. Biologists pulling and spraying weeds attracted bird ticks and then unwittingly carried them back to camp. Soft-bodied seabird ticks look like buffy, crenulated grains of coral sand. Mobile and heat-seeking, they are hard to

cate all the sandbur from Laysan, especially if there was a massive seed bank in the soil. But the persistent weeding paid off. In 1998, there were virtually no sandbur seedlings to show to a new team of weeders. The sandbur seedbank proved "bankrupt," and the native grasses are coming back strong (too strong for the federally endangered Laysan Sedge; the species was down to three plants in 1994, but thanks to weeding away the bunchgrass, the sedge has increased to over a hundred plants).

With the weeds under control, further restoration is planned. The island that Hugo Schauinsland once knew is the goal of the future restoration detect as they insidiously work their way under loose clothing to suck your blood. It is impossible to avoid them. Reactions vary among people, from slight skin irritation to intolerably itching bites that easily become infected and eventually develop into hard welts. Even noninfected bites take months to heal fully.

I seemed especially sensitive when, one day, my tick bites itched fiendishly and became bloody lymph pustules. Dehydrated, with a headache and a sore back, I sought shelter from the sun only to find that my hot tent had been overrun with tiny red ants (one of fifteen or so species introduced to the island). They had moved in with their millions of eggs and many queens. After moving my tent away from the grass and onto the beach, I was finally free of ants. But it slowly dawned on me that the tick bites on my swollen ankles were causing a fever. I was sweating one minute, chilly the next. My appetite was off and my dreams had a purgative quality. In several days my aching joints and listlessness passed and I returned to the weed war. Soon others came down with a fever and each, in turn, took a couple of days to recuperate.

I did not realize in 1991 that we were the first major wave in an undescribed illness. Subsequent teams of weeders experienced similar symptoms, some worse (one woman had to be medevaced from Laysan because of a persistent fever), others not at all. The USFWS, fearing a public health threat to their squads of weed-pulling volunteers, initiated an inquiry. In 1994, Cedric Yoshimoto, an epidemiologist from the University of Hawai'i's School of Public Health, considered the evidence and reported: "Since 1990, wildlife management personnel in field camps on several of these islands (particularly Laysan Island) have become ill, sometimes requiring medical treatment via radio and once requiring dispatch of a ship for evacuation. A retrospective survey and subsequent prospective surveillance has identified a newly-

described illness of humans termed 'Laysan fever (LF).' It is associated with bites of the seabird tick complex *Ornithodoros capensis* and occurs primarily between the months of May and August. . . . Laysan fever joins a short list of human illnesses associated with seabird colonies in various parts of the world."[6]

I began to wonder if this was a new disease or one that had been overlooked in the past. After all, biologists had been working out here for decades without getting tick fever. In the 1960s, the Pacific Ocean Biological Survey Program (POBSP) put many biologists in the field to help survey over 4 million square miles of the central Pacific Ocean (out of a total 65,000,000 square miles of the whole Pacific Ocean).[7] Never before had such a vast survey been undertaken over such a lengthy period (1962–1972). They visited every island and atoll without experiencing, or at least reporting, a pervasive illness.

But during the POBSP period, something changed in the environment. The Army tested germ warfare agents in the central Pacific Ocean.[8] In fact, the POBSP, sponsored by the Smithsonian Institution, was paid for by the U.S. Army. The seabird project was modest at first but eventually became a $2.8 million dollar contract, initially administered by the Fort Detrick Laboratory in Frederick, Maryland. The military post was well known as the Army's chemical warfare center; biological warfare was not yet officially invented. The POBSP collected duplicate sets of seabird distribution data; one set went to the Army, the other was put into the public record at the Smithsonian Institution (no longer the poor useless, innocent, harmless old fossil that Mark Twain described in *Innocents Abroad*). However, the Army's portion of the project was top secret. The POBSP biologists collected seabird blood samples and ticks, sending them back to Fort Detrick and Deseret Test Center, Utah, for analysis.[9]

Eugene Kridler, the manager of the Northwestern Hawaiian Islands National Wildlife Refuge, recalled that in 1964 he found out that the POBSP crew on Laysan were taking too many blood samples from the birds, with considerable loss of birds. But Kridler was stuck in Honolulu with no way of checking what they were really doing. Philip Humphrey described the purpose of the POBSP in the 1965 Annual Report of the Smithsonian Institution: "Perhaps the most important practical accomplishment of the Smithsonian survey will be a delineation of the environment over a relatively short period of time. This will provide a baseline of comparison for biologists concerned, 10 or 20 years from now, with measuring the effects of man-made modifications of the environment on natural populations of organisms. The need for such a baseline is most urgent today, when man, in his struggle to advance himself, is changing the face of the earth at an appallingly rapid rate, and is subjecting the total environment—water, atmosphere, and living tissues—to physical and chemical influences which need to be measured now and in the future. For unless these fundamental changes in his environment are properly assessed, man himself, through ignorance, may fall victim to his own progress."[10]

To one from the "conspiracy theory generation," the phrase *living tissues* is a curious word choice. It seemed to me that Dr. Humphrey was hinting that something nefarious was going on, something that could affect the metabolic makeup of organisms. In 1969, an article in *Science* reported that Dr. Humphrey was "sure" the Army wanted to test chemical and biological weapons in the Pacific and was looking at the findings of the ecological surveys to ascertain whether a potential site was "safe." He speculated

that two biological agents were possibly tested, Venezuelan equine encephalitis and Q fever. He also said that the POBSP biologists were given a series of inoculations to protect their own living tissues. In fact, in 1964, two bacteriological agents were released near Johnston Atoll. Tularemia and Q fever were aerially dispersed in twenty trials to test the virulence of the agents on Rhesus monkeys.[11]

I began to wonder if it was possible for the fevers to have persisted in the environment since the mid-1960s. Was one now infecting some people on Laysan Island? It seemed improbable, but I needed to learn more about the diseases before I could draw any conclusions. I started searching through the Army's studies, but many were classified. In the few unclassified documents that were available, I found two incriminating abstracts: "Susceptibility of Sooty Terns to Venezuelan equine encephalitis (VEE) virus"[12] and "The susceptibility of birds to tularemia: The Wedge-tailed Shearwater and Black-footed Albatross."[13]

Upon further investigation, I learned that these papers had been destroyed in 1969, when

Black-footed Albatross fledgling exercising among glass balls.

the United States formally ceased to conduct offensive biological warfare. Also, I learned that these studies were not conducted in the field but in a laboratory. Yet, I could not imagine that tropical seabirds, with their need for seafood and seawater, would survive trips to laboratories in Utah and Maryland, let alone provide meaningful data.

I turned my attention to Q fever. A U.S. congressional inquiry into offensive chemical and biological warfare testing described Q fever as a disease of animals. The main sources of infection to humans are sheep, goats, and cattle; transmission is most frequently by air. Q fever is caused by a rickettsia, defined as any of several bacterialike microorganisms of the genus *Rickettsia*, parasitic on arthropods and pathogenic to humans and animals (named after Howard T. Ricketts, an American pathologist [1871–1910]). Q fever rickettsiae are extraordinarily resistant to environmental factors such as temperature and humidity. Very large amounts can be produced in embryonated chicken egg (20,000 million microorganisms per milliliter) and can be stored for a long period of time. It was chilling to learn that: "New natural foci, in which infection may persist for many years, may be established after an aerosol or other type of bacteriological (biological) attack. . . . Arthropods (insects, ticks) also play an important part, along with other creatures, in the maintenance of pathogenic agents in natural foci. A man exposed to a natural focus risks infection, particularly from arthropods, which feed on more than one species of host. A bacteriological (biological) attack might lead to the creation of multiple and densely distributed foci of infection from which, if ecological conditions were favourable, natural foci might develop in regions where they had previously never existed. . . ."[14]

Did the infective agents persist in the environment and infect birds throughout the Pacific Ocean? Were ticks infected with Q fever spreading this "new" disease to people? Was Q fever, or a

form of it, now showing up on Laysan Island after 25 years? Epidemiologist Yoshimoto wrote: "Infectious agents are high on the list as suspects for the etiological agent(s) of Laysan Fever. Some agents may be present in the environment. For example, *Coxiella burnetii*, the agent of Q fever may persist in the soil and water for months, may infect some wild and domestic birds, may infect a variety of tick species, and may persist in tick feces for well over a year. . . . The symptoms of Q fever overlap with those of LF to a considerable extent."[15]

The extensive role of birds as tick hosts is generally unappreciated by biologists and epidemiologists alike. Owing to their extensive migrations and to their high rate of tick infestation, birds have an important role in distributing ticks that also carry infectious diseases.[16] But in this instance, I was reassured by Dr. Thierry Work, a USFWS pathologist, who reported that bacterial rickettsiae were never found in the human blood samples taken from people on Laysan Island. "The clinical signs of Laysan Fever are non-specific. That is, they could just as easily apply to a panoply of zoonotic diseases (diseases transmitted from animals to people) including Q fever. Thus, diagnosing someone as having Q fever based on clinical signs alone is inappropriate. There have been several viruses isolated from *Ornithodoros capensis* worldwide. At least two of these have been documented from Oahu (Midway virus) and Johnston Atoll (Johnston Atoll virus). More recently (1993), I submitted ticks collected from Midway to the Centers for Disease Control and they isolated a virus there (a Hughes virus). So, as you can see, there are a lot of viruses running around out there naturally that could all be potential candidates as the causative agent of 'Laysan fever.' "[17]

Since 1996, there have been fewer cases of Laysan fever reported and it appears that it is declining in the environment. However, it remains unknown whether the disease is caused by the body's reaction to tick bites or by an infectious

Black-footed Albatross "watching TV" on Laysan Island.

microbe transmitted by ticks. So perhaps it is out-landish to suggest that Q fever, sprayed across the Pacific by the U.S. Army, infected us on Laysan 30 years later. But the information gleaned during the inquiry into the secret side of POBSP suggests that something occurred somewhere in the central Pacific—something the government wishes to keep hidden.

Around the time that biologists started coming down with fevers, a more ominous situation was discovered. Any lingering illusions I had about the pristine nature of Laysan Island vanished. Biologists noticed an area above the high-tide line where many animals were dead. Albatross chicks dropped dead. Ghost crabs eating their carcasses

became virtual ghosts. Blow fly maggots eating the crabs died. What was it? Animal, vegetable, or mineral? Nuclear, chemical, or biological? Name your poison! The problem proved to be carbofuran, a pesticide, that evidently floated to shore in a container, cracked open, and spilled its contents. The "dead zone" remains off limits today.

Unfortunately, wildlife in this remote region is affected by the pervasive pesticides in our environment. Pesticides like PCBs, dioxins, and furans have filtered their way into the marine food webs from terrestrial sources. Seabirds, as predators, concentrate the long-lived toxins. Albatross eggs show toxic levels of contamination. Egg death occurs in 2 to 3 percent and is greater in Black-footed Alba-

Plastic creatures collected from the flotsam on Laysan Island.

reached Hawai'i. What a fantastic world that would be!

Plastic rubbish jettisoned overboard from fishing boats affects seabirds in unexpected ways. While studying the food habits of Hawai'i's seabirds, Craig Harrison found that many gut samples contained bits of plastic.[19] Seabirds mistake floating plastic pieces for marine creatures and ingest them. The birds feed their chicks via regurgitation, passing plastics on to the chicks. Plastic takes up space in the bird's stomach that food could occupy, sometimes causing malnutrition or lacerating the gullet of albatrosses. The indigestible material is periodically spewed forth. I once found a rubber glove regurgitated by an albatross. All too often plastic kills the birds. The interior of Laysan Island is littered with toys, bottle caps, and other plastic trash carried inland by albatrosses and scattered by the wind blowing away the carcasses of dead chicks. In addition, plastic fishing floats, shower slippers, sake bottles, rope, light bulbs, crates, netting—you

trosses than in Laysan Albatrosses. The rate of birth defects such as crossed bills is 1 in 14,000 for Laysans and 1 in 7,800 for black-foots. Cross-billed hatchlings were not reported in these birds until the late 1970s in spite of intensive studies between 1957 and 1972. Yet current exposure levels will probably not harm the overall population health of albatrosses.[18]

Times have indeed changed on Laysan. Eric Schlemmer recounted the good old days and how his family loved living on Laysan. He remembered Christmas of 1906 when his sisters were sewing button eyes on their rag dolls and how Max accidentally set the "Christmas tree" *(naupaka)* on fire. Everyone had a good time with their simple gifts because they knew no other way. Today, the flotsam alone would bring all the toys any kids would want. Plastic monsters, army men, pink tuba players, bizarre superheros: all regularly wash ashore. It's as if a giant tsunami sucked up Japanese children's small, plastic toys and spewed them out on Laysan. Imagine if these plastic creatures could evolve like the original organic material that

Nama sandwicensis grows in a plastic bleach bottle.

The restoration of Laysan Island will include the introduction of Nihoa Palms to take the place of Laysan's now extinct native palms, seen here in the 1880s. (Bishop Museum photo)

name it—washes ashore and blocks the upper beach zone where Black-footed Albatrosses breed. Ultraviolet light exposure eventually breaks down the plastic into minute chips that some seabirds mistake for food.

Large pieces of fishing nets can be lethal to monk seals. Fish traps, trawl nets, purse seine nets, and open-ocean drift nets all continue to kill marine life after they are lost at sea. Since 1982, there have been 155 seals observed entangled in discarded fishing gear. Some seals were hooked and at least one had drowned. In 1998, an all-out effort to remove marine debris was launched by a multi-agency task force that included USFWS, the National Marine Fisheries Service, the Coast Guard, and several other nonprofit organizations. Over 6 tons of ropes and netting were removed from the beaches and reefs of the wildlife refuge in a week. The nets, some weighing up to 1,500 pounds, were collected for disposal and analyzed to determine the country of origin. Much of the gear was from Asia, but a substantial portion was from the U.S. fisheries.[20] Violations can cost up to $25,000 a day, but enforcement is virtually impossible.

Among the debris there is the occasional happy message in a bottle. My best friend found one on Laysan: "To whom it may concern; Come on over to my house and if can not please write back soon if you can, we might go fishing because it is fishing season, we might play football, and baseball and eat cookies maybe. Sincerely, Noel D. Kingsford, Coos Bay, Oregon, 23 April 1973." My friend answered Noel's letter 7 years later when the note was found, asking if he had any cookies left. No one replied. Most notes in bottles are illegible because of exposure to sun and/or moisture. The more clever bottle launchers wrap their notes in aluminum foil to protect them from the sun and seal the bottles well.

Like parents whose teenagers are out late on a Saturday night, the USFWS managers of Laysan Island anticipate a call in the night saying that something has happened, and periodically something does. In 1969 and again in 1976, foreign fishing vessels ran aground on Laysan. Rats were evidently on board and if they had gotten ashore, their attempted eradication would have made the weed war look like a picnic. "Ratspills" must be prevented at all costs. These and other hazards of the twenty-first century, such as global warming, increasing ultraviolet light, oil and chemical spills, lost fishing gear, and other debris from the sea, will forever threaten the "gem of the Leewards," lovely Laysan Island.

References

1. Lipman (1984).
2. U.S. Department of the Interior (1984).
3. Newman (1988).
4. King (1973).
5. Marks (1995b).
6. Dr. Cedric Yoshimoto, 1994, personal communication.
7. E. H. Bryan Jr. (1942).
8. Gup (1985).
9. Boffey (1969).
10. Humphrey (1965:30).
11. Regis (1999:259).
12. Miller et al. (1963)
13. Brown and Cabelli (1964).
14. United Nations, Group of Consultant Experts on Chemical and Bacteriological (Biological)Weapons (1970:128–129).
15. Dr. Cedric Yoshimoto, 1994, personal communication.
16. Hoogstraal (1972).
17. U.S. Fish and Wildlife Service pathologist Thierry Work, 1998, personal communication.
18. Ludwig et al. (1996).
19. Harrison (1990).
20. TenBruggencate (1998).

16

Lisianski Island

Lele ka ʻiwa mālie kai koʻo.

"When the frigatebird flies, the rough water will be calm."

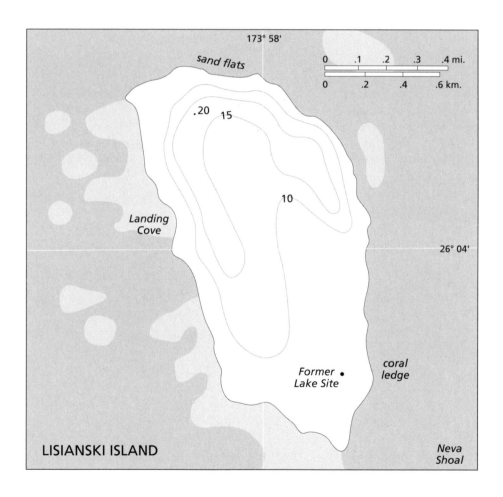

Map of Lisianski Island.

About 1,000 miles northwest of Honolulu and 115 miles west of Laysan lies Lisianski Island. This third largest Northwestern Hawaiian Island is less than half the size of Laysan but shares a similar geology. About 20 million years ago, in the Miocene Period, geologic forces raised the tip of a vast underwater coral bank above sea level. Today, about 400 acres of dry land composes Lisianski Island. The highest point is a sand dune 40 feet high in the northeast portion of the island; the lowest point is a depressed area in the southern half of the island that has a channel-like formation leading to the ocean.[1] A wide beach surrounds Lisianski's interior, which is a savanna of dry grass, rather than the open water of Laysan's lake. Described as a "flat and uninviting looking islet," a few Tree Heliotrope shrubs, some dead Ironwood trees, and one live Pisonia tree dot the horizon on this parallelogram-shaped island.[2]

Although it may be uninviting to humans, the island is home to about three-fourths of all the Bonin Petrels nesting in Hawai'i. Under the cover of darkness, over a million of these tubenoses return to dig their burrows under the bunchgrass in the center of the island. Bonin Petrels have a sweet disposition, especially when compared with their neighbors, the Wedge-tailed Shearwaters. The Bonin Petrel has a soft, purring cry; the shearwater works itself into a caterwauling fit that sounds like a colicky baby. Both nocturnal species nest underground, but the larger, more aggressive shearwaters forcibly evict the gentle petrels from their burrows.

Lisianski Island is rich in biodiversity because of its proximity to major coral reef ecosystems. The island is surrounded by Neva Shoal, a reef that extends to the southeast for 106 square miles. Another major reef system, Pioneer Bank, lies 22 miles to the east.[3] These cartographic names honor the island's first recorded visitor, Iurri Fedorovich Lisianskii (aka Urey Lisiansky), captain of the ship *Neva*. Captain Lisiansky, in tandem with

Captain Krusenstern of the ship *Nadeshda,* were the first Russians to circumnavigate the globe. The ships parted company after rounding Cape Horn, and Lisiansky then sailed into the Pacific and visited Easter Island in April 1804. After passing through the Marquesas Islands, he arrived in the Hawaiian Islands and voyaged around Kaua'i and Ni'ihau. On 20 June the *Neva* headed north and wintered over in Kodiak, Alaska, presumably trading for sea otter pelts. Lisiansky then headed for Sitka, the Russian capital in Alaska, on 20 August 1805, before setting sail to rendezvous with Krusenstern in Macao. Sailing north and west of the Hawaiian Islands, the crew of the *Neva* sighted flocks of seabirds and set a special sea watch on 15 October. They saw no sign of land and relaxed their vigil. All too soon, for later that night, the vessel hit a reef.[4]

Lisiansky ordered the heaviest objects, such as cannons, jettisoned in shallow water from which they could be retrieved. The *Neva* floated off the rocks and appeared out of trouble when a sudden squall drove the ship on the reef again at 5 A.M. Removing the rest of the heavy anchors, chain, and cables, the crew managed to extricate themselves. At daybreak, they sighted a sand isle and landed on 18 October. The Russians found a beached redwood log, 21 feet in circumference. More important, Lisiansky found a Hawaiian artifact, a gourd calabash on the beach.[5] The crew slaughtered several seals for fresh meat and quickly departed. Sir Peter Buck, author of *Explorers of the Pacific* and former director of the Bishop Museum, speculated that Lisiansky hastened back to Russia in advance of his companion Krusenstern to be first to receive the honors for their mutual accomplishment of circumnavigating the globe.[6] After his early retirement at age 36, Lisiansky wrote in his memoirs: "To the southeast point of the bank where the vessel grounded, I gave the name of *Neva*; while the island itself, in compliance with the unanimous wishes of my ship's company, received the appellation of Lisiansky. . . . On landing, we were much annoyed by the birds, many of which made their attack flying, while others ran after us, pecking at our legs: it was with great difficulty we could keep them off, even with our canes. . . . The heat of the day was excessive, and, almost at every step, we sunk up to our knees in holes, that were concealed by overgrown creeping plants, and contained nests, as we supposed, of various birds; for we often heard their cries under our feet from being trampled upon. . . . [Lisianski Island] promises nothing to the adventurous voyager but certain danger in the first instance, and almost unavoidable destruction. . . . I cannot help feeling grateful to Providence, persuaded that, without his aid, like Mr. de la Pérouse and companions, not one of us would ever again have beheld his native land."[7]

The next ship to visit was the Russian ship *Moller,* in 1828. The on-board surgeon, C. Isenbeck, explored the island and collected birds and was the first to describe a species of duck from the island. In 1844, an American whaling ship, *Holder Borden,* under the command of Captain Javes J. Pell,

wrecked on the reef. An account in the newspaper *The Polynesian* read:

The *Holder Borden* was a new and beautiful ship of 442 tons, one of the most valuable of the whaling fleet and sailed from home, at the expense of $65,000. She was insured, and it is supposed the cargo was also. . . . On April 5, 1844 having on board 800 bbl. [barrels] sperm and 700 bbl. black fish oil, she ran lightly bows on, upon a bank of sand, and shortly after the stern swung round and struck a coral reef, from which it was found impossible with their utmost efforts to haul her off and where she thumped until she bilged.

The island was found to be 174°55′W & 26°01′N. On Turner's map of the world it is laid down as Drake's Island. It is about three miles in circumference and thirty feet in its greatest elevation with a swampy lagoon covered with grass in the center into which the highest tides partially flow. Beach grass and a few flowering shrubs are the only vegetation. Fresh water, though somewhat brackish is abundant.

Hair seals, turtles, wild ducks and other fowl are plentiful so that the crew, 36 in number, besides the provisions saved from the ship, were abundantly supplied with fresh food. The wild ducks were readily tamed. Potatoes, melons and other fruits and vegetables were planted and came up but withered for want of water, there being but 12 hours of rain in five months. The heat was great, the thermometer ranging from 94–98° and sea fowl, it is said, frequently near the well suddenly dropped down dead, the effect it appeared to be of

Biologist Craig S. Harrison huddles for shade under the tarp tied to the refuge sign at Lisianski Island.

overpowering heat. . . . [The men were] obliged to make saws from hoop-iron, but having a carpenter and blacksmith with a forge and coal, they persevered and by Sept. 8, they completed a vessel (35 ton) of a good model, painted, sheathed and copper fastened throughout—named HOPE.[8]

Hope set forth from Lisianski Island with eleven men on 14 September 1844, reaching Honolulu on 8 October. Pell returned to Lisianski to save his remaining men and salvage the *Holder Borden* and her valuable stores of whale oil. As a parting gesture, Pell planted about eighty coconuts, knowing how useful they would be for the next marooned crew. He also inadvertently introduced House Mice, which were later noted in abundance.

On 24 May 1846, the whaler *Konohasset* wrecked on the reef near Pell's Island, as it was called on mid-1800 charts. (The charts of island positions were coveted by traders and exact locations were purposefully misrepresented, especially if valuable resources, such as guano, were to be found on them.) Within a month, a 22-foot boat was

Masked Boobies on old redwood log at Lisianski Island. This log is much smaller than the one Lisiansky himself found here in 1804, which measured 21 feet in circumference.

crafted from the wreck and set forth for help. Forty-two days later, the *Konohasset Jr.* reached Honolulu and aid was sent to the survivors on Pell's Island.[9]

In 1857, the schooner *Manuokawai,* with Captain John Paty at the helm, visited the Leewards to annex them for the Hawaiian Kingdom. On 3 May Paty landed at "Liseanskey's Island" (it took many years before the spelling of the island's name became uniform) and took possession of the island in the name of the king. Paty discovered the presence of guano in what he considered to be commercial quantities. He also found three casks, a broken spar that had been used as a "crows nest," and a cookhouse with the name *Holder Borden* carved in the wall. He surmised that *Holder Borden* was wrecked on another island about 1 degree west: "I lost three days of time in looking after it, and can safely say that Pell's Island does not exist in this ocean. The 40 domesticated ducks which Capt. Pell speaks of, must have resumed their roving propensity, as I did not see the sign of one on the island. I have understood that Captain Pell planted some cocoanuts on the island in 1844—not a sign of them exists now in 1857, or any other vegetation, except coarse grass and a few small running vines. I planted a handful of white beans, and a few sweet and Irish Potatoes. . . . There is what has been a lagoon near the southern part of the island, in the centre of which fresh water was found by digging 5 feet. Birds, fish, seal and turtle abound here, but not so plentiful as at Laysan Island."[10]

On 24 July 1872, the brig *Kamehameha V* made it through the maze of reefs at Neva Shoal to discover that another ship's crew had been less skilled or lucky. *The Friend*

Red-footed Boobies nesting. The male lands on the female, presents nesting material, and mates. He flies off to find or pilfer more material and repeats the process.

carried portions of Captain E. Wood's account of his cruise:

July 24th, made the reefs at Lisiansky Island, and saw a wreck on the reef to the S.E. of the island. . . . On the west side found a studdingsail boom; rigged as a flagpole, with signal halyards rove. On the N.E. side found the long boat on the beach, having drifted ashore. She had been rigged for sea, and been capsized or stove on the rocks. She had a canvas deck, a bowsprit, rigged the mast with the rigging [that] had been cut clear of the boat. She was badly stove. On the south end of the island found the place where the crew had landed. There was found her quarter boat, with a mast and remains of a sail, moored to two water casks, half full of fresh water, and a grapnell off shore. She was a wreck, being badly stove. On the sand beach were the remains of clothing, some carpenter's tools, a box of bread, a box of Jenny Lind cakes, and three of soda crackers, all wet with rain water; a tin chart case, empty; some tins of pine apple, put up in New York; the poles that had been used for a tent, a topmast studdingsail, made up. . . . Found the [main] wreck to be a brig, laying with her head to the N.W., full of water; the larboard rail out of water; the main topmast gone at the cap; main yard across, with the remains of mainsail; the foremast, topmast and topgallant standing with all the yards a cross; foretopsail set; remains of topgallant sail flying, which was what had appeared like a flag at half mast the day before. Two casks of fresh water were lashed securely on the larboard quarter outside. She lays on the N.E. side of the reef, which extends ten or twelve miles to the S.E. No breakers in sight from the wreck. Mr. Cahill cut a hole in the house on deck, and getting into a state room, found the vessel's log-book, which he brought away. By this it appears that the wreck was that of the North German brig *Wanderer,* of Hamburg, from San Francisco bound to Port May on the coast of Tartary.[11]

No sign of the crew was ever found, and the mystery of the *Wanderer* was never solved. The language and expertise of the sailing days seems exotic and obtuse to us today, but the skill level and years of knowledge remind us how much is lost to the ages, including the means to survive on a desert island. It is incomprehensible that any of us bird biologists who ever camped on Lisianski, huddled like hermit crabs under a plastic tarp tied to the refuge sign, could have survived a few weeks without outside support. Digging a well and sucking seabird eggs was the extent of our survival skills. The hardiness and skill possessed by the lowliest sailor of bygone times surpassed all our collective knowledge and experience. I am afraid to say that we would not have had the courage to set forth from Lisianski in the first place and if we did, it would have been more than a miracle if we ever saw land again.

Such a miracle saved the crew of the bark *Afton,* wrecked on Lisianski in 1887. Captain Gilmour and crew manned the lifeboats and set sail for Honolulu, only to find that they could not make any headway against the trades. So they decided to run west with the wind. A month and 3,000 miles later they reached Guam, without the first mate, who had fallen overboard.[12]

The guano discovered by Paty in 1857 was sought after by the North Pacific Phosphate and Fertilizer Company, which leased Lisianski and Laysan Islands. The company concentrated its efforts on Laysan and evidently never built anything on Lisianski. Max Schlemmer tried to shift his operation here after the guano on Laysan had been expended in 1904, but his application to lease the island was turned down repeatedly. Not until early 1909 did Schlemmer get permission to lease Lisianski (and Laysan) for $25 a year.[13]

Although the Hawaiian guano lost out to Chilean nitrates for market share, the French millinery trade advanced and attracted Japanese feather hunters to these islands. After exploiting their own seabird rookeries, the Japanese raided the Northwestern Hawaiian Islands. In early 1904, thirty-eight Japanese were dropped off on Lisianski to collect feathers. They worked for at least 10 days before their ship was wrecked on the reef. The entire crew was hard at work slaughtering birds when Captain Nilblack of the USS *Iroquois* made a surprise visit on 11 April. He attempted to inform the Japanese of their trespass, but he spoke no Japanese and they spoke no English. However, he did alert the revenue cutter *Thetis* about the situation. The *Thetis* reached Lisianski on 16 June 1904 and evicted the Japanese from the island, evidently not against their will because they were on starvation rations. They had plenty of birds to eat, however, because they had killed an estimated 284,000 and collected 110 sacks of wings, 100 crates of whole birds, and 116 cases of wings and birds that were left abandoned on Lisianski. Apparently, a shipload of feathers was removed by the Japanese before the raid, but the ship ran aground at Pearl and Hermes Reef and lost its cargo.[14]

Captain Weisbarth, a potential partner of Max Schlemmer, went to salvage the abandoned bird skins on Lisianski, intending to sell them in British Columbia and get rich. Distancing himself from Weisbarth, Schlemmer denied ever having charge over the Japanese or of his alleged partner Weisbarth. Schlemmer said that he might have become a partner in the venture if Weisbarth had obeyed his orders, but since he "preferred to hunt fortune on his own hook . . . he'll have to play in his own yard after this. 'I knew the Japanese were on Lisiansky,' said Schlemmer last night, 'and I protested and had the *Thetis* sent out after them. I was in charge of the island for the Fertilizer Co. and my orders were to protect the birds. . . . I don't think he is going to get rich off those skins. Skins must be kept dry and after lying on the island all these months I don't think they will be in condition to sell.' "[15]

Actually, Schlemmer had visited Japan in the early 1900s, and he drafted a contract in 1908 that granted the Japanese whatever rights he held to Laysan and Lisianski in return for $150 a month in gold. The Japanese returned to plunder Lisianski in 1909 and made off with fifty bales of bird wings. Rumors of poaching reached Honolulu and again the *Thetis* was dispatched to round up the perpetrators. When they reached Lisianski on 19 January 1910, an armed landing party arrested the poachers and destroyed about one and a quarter tons of feathers representing 140,400 birds worth about $97,000. At Lisianski alone, over a million birds perished in the feather trade. The ensuing public uproar in early 1909 prompted President Theodore Roosevelt, in the final days of his administration, to establish the Hawaiian Island Bird Reservation by executive order.

Protection from feather pirates was one thing, but the release of mice and rabbits on Lisianski around the same time destroyed the island's ecology. In 1893, John Cameron, captain of the ship *Ebon*, ventured into the Leewards to collect seals, turtles, and sharks. He characterized the islands as "Lisiansky of the Mice, Laysan of the Birds!"[16] His journal includes a description of Lisianski Island as House Mouse heaven:

We settled ourselves for an appetizing supper of fresh food when myriad's of mice attacked our meal ravenously and utterly without fear. Drive them away we could not; we slaughtered them by the hundreds, yet they would not be denied. A full hour elapsed before we could eat in some semblance of peace; then each of us had to hold his food in one hand and a stick in the other. During the night the pests continually galloped over us; they did not, however bite us, though that seems remarkable, since there was little on the island for them to eat, unless they devoured one another.

By day we were not molested. In preparation for the certain onslaught of the next evening we brought the ship's cat ashore. Surely he would protect us. Not so. As the gloaming fell the mice swarmed upon us in numbers exceeding those of the previous night. Some mouse runner with a fiery cross must have dashed about the island to summon the clans. Our cat began to kill in a feline paradise: here were all the dreams of the cat family come true: mice in impossible numbers, mice of incredible boldness. In a whirlwind of activity Tom slaughtered, too engrossed to bother torturing his tiny victims. Yet the armies thronged too rapidly for him. Over his face crept disgust, dismay, fear as of the supernatural. Soon he sat himself on his haunches and stared at us; surrounded by wind-rows of dead mice, he let the wee animals run unharmed between his paws. The beasties had conquered: we removed our camp to the beach and dug a moat, which soon filled with sea water, as a protection against them. Only one succeeded in crossing the ditch. It was pointed out to Tom, but he would not touch it.[17]

The mice had a field day until the rabbits arrived, possibly introduced from Laysan by Max Schlemmer.[18] Before the rabbits ate themselves out of house and home, an attempt was made to introduce Laysan Rails. As mentioned in chapter 14, the dire plight of the Laysan Rail was due to the destruction of Laysan Island by rabbits. Efforts to find safe havens for the species were a priority before it disappeared completely. In 1913, Alfred Bailey and George Willett, members of the Bureau of Biological Survey, introduced forty-five rails to Lisianski Island.[19] Unfortunately, by 1915, Lisianski was denuded. Carl Elschner visited the island and wrote in the newspaper *The Honolulu Advertiser:* "Save for the presence of many thousands of seabirds, the island presents such a dreary and desolate appearance as has never confronted me even on my travels over the deserts of North Africa and Syria. Walking over this island is extremely difficult on account of the many holes in

which the birds nest. Without doubt the avifauna is poorer in species than on Laysan Island. The rabbits introduced have just exterminated the flora, probably never very luxuriant, with the exception of tobacco (brought by Capt. Schlemmer); it is almost impossible to realize the complete extermination of the vegetation; now the rest of these rabbits (we found many dead but very few living ones) will have to submit to starvation, certainly a terrible fate for a once large population of these poor doomed creatures."[20]

Apparently the rabbits ate themselves into oblivion by 1916, taking the rails and mice with them.[21] Environmental destruction also destabilized the dunes and windblown sand may have filled the low depression that was formerly used by ducks. The marooned sailors from the *Holder Borden* reported that wild ducks were plentiful in 1844 and evidently ate a bellyful because ducks were never recorded here again. An intriguing biological mystery surrounds the reports of ducks on Lisianski Island that may shed some light on Hawaiian sailors traveling through these far western islands.

Alan C. Ziegler carried out archaeological and paleontological research on Lisianski in 1990. He excavated a site in the low-lying area of Lisianski and discovered dark, blackish brown soil that appeared to be the accumulations of marsh plants. He also found thirty-nine bones of ducks, including one from a bird evidently too young to fly, proving that "Laysan" Ducks once bred on Lisianski Island.[22]

One rail bone, which was probably from the introduced Laysan Rails, was also discovered. "That this bone was found at all suggests that the sampling effort was probably sufficient to have revealed an indigenous rail had there actually been one."[23] Evidently, the flightless rail was endemic to Laysan, but the duck had a much wider distribution than previously thought. In fact, skeletons of this duck, including juveniles, have been recovered recently from a variety of prehistoric sites throughout the main Hawaiian Islands. "The Laysan Duck may have been widespread in the main archipelago and once lived even in forested situations at fairly high elevations, far from water. Because of such adaptability, a population perhaps could have existed on Nihoa as well. That there was no apparent differentiation, at least osteologically, in the population from Laysan and Lisianski, compared to the main islands, suggests that the Northwestern populations either were of recent origin or were supplemented fairly regularly by immigrants from the main island."[24]

It seems odd that a duck on its way to flightlessness at Laysan Island could have traveled over 900 miles to this region fairly regularly from the main islands. I would like to suggest that it is at least possible that ancient Hawaiians carried ducks to these islands as a food source for future visits. It was a well-known practice to cache food in places that might be revisited. Voyaging Polynesians also traveled with livestock in the form of pigs, dogs, rats, and chickens. Why not ducks if they were so common in the main islands and "easily tamed," as *The Polynesian* reported in 1844?

There are tantalizing tidbits to support the theory that Hawaiians ventured this far west. Captain Lisiansky reported in the ship's log on 18 October 1805 that he had "found on the beach a small calabash, which had a round hole cut on one side of it. This could not have been drifted from a great distance, as it was fresh and in good preservation."[25] This suggests that a water container could have been left on the island and that Lisiansky was not the first discoverer of the island. Archaeological research carried out by Dr. Ziegler initially suggested prehistoric Hawaiian visitations. Ziegler found an "enigmatic unworn flake-like fragment of fine-grained basalt. . . . A less conservative explanation for the presence of the basalt flake would be that it was a fragment from a prehistoric Hawaiian adz or other basalt tool brought to the island. . . ."[26] But subsequent analysis suggests the basalt may be Japanese. However, it remains a possibility that in addition to the gourd calabash and basalt flake, Laysan Duck distribution may also be evidence of Hawaiian voyaging prowess.

Perhaps the best navigators in the world visit Lisianski every year. It is one of the few places where Bristle-thighed Curlews perch atop green glass balls—their spyglass—for a better view. Curlews are relatively rare both in the Hawaiian Islands and in North America. Somewhere between five thousand and ten thousand curlews breed on the Alaskan tundra around Nome and the Yukon River delta. In late summer, they gather into small flocks and migrate to remote islands in the tropical Pacific Ocean. As the species name *tahitiensis* implies, the birds are seasonal Polynesians, wintering over some 50 degrees of latitude from Midway in the north to Pitcairn Island in the south. Called *kioea* (pronounced kee-oh-ey-ah, its onomatopoeic Hawaiian name) after their easily imitated whistle, several hundred overwinter in the Northwestern Hawaiian Islands and a few even spend the entire year.

Bristle-thighed Curlew bathing in a tide pool at Laysan Island.

The unique bristles protruding from the feathers on the curlews' thighs are thought to be tactile organs used in courtship, though no one knows for sure. Indeed, little was known about this curlew until recently. Their breeding grounds in remote western Alaska were only discovered on 12 June 1948. Research on Laysan Island by Jeff Marks has provided some new insights into this large shorebird. One individual, making its first flight south from Alaska, probably had not been on the island for a week before it was captured by Dr. Marks. The bird was banded and remained on Laysan throughout 1990 and 1991. It was still present in 1993 and probably left for Alaska on 2 May, during the week that most curlews depart Hawai'i for Alaska.[27]

This bird's presence suggests that some juvenile curlews remain in Hawai'i until at least the beginning of their third year, though some do not breed until 4 or 5 years old. Breeders migrate non-stop to Alaska. The urge to migrate is so strong that once a broken-winged curlew walked the entire length of Laysan Island to arrive at the staging area where the island's population of curlews met for their flight north. That seems so pitiful, but life on the island is not that bad foodwise (options are stealing eggs, bashing crabs, and eating roaches); besides, the other curlews will be back in a short 3 months for their 9-month vacation in Hawai'i.[28]

The curlew is a relatively trusting bird and may even approach out of curiosity, especially if its "wolf whistle" *whee wheeo* is imitated—a technique Marks has used to draw birds close enough to read their bands. For several years he has followed the movements of banded curlews and was astounded to observe some of them using tools. Tool use in birds is extremely rare and is limited to a few bird species. Like Egyptian Vultures, Bristle-thighed Curlews drop rocks on eggs to crack them open and expose the yolk for drinking. Curlews only use rocks to open large albatross eggs; smaller eggs are easily picked up and slammed to the ground or skewered. Because of their appetite for eggs, curlews are often mobbed by Sooty Terns. Curlews also eat a variety of insects and crustaceans captured with their probing bill, the tip of which is sensitive to movement and vibrations in the ground.[29]

In the wave-sculpted, coral limestone slabs on Lisianski's eastern shore, I once pursued a Ruddy Turnstone with a broken wing. I thought I had my photographic subject cornered when it reached the edge of the last slab. As I raised my camera, the turnstone jumped into the water and immediately disappeared. A giant *ulua* had sucked down the turnstone instantly. Like the *paparazzi*, I had driven the poor beautiful bird to its death!

Ruddy Turnstones, *'akekeke* in Hawaiian, are the most numerous shorebirds in the Northwestern Hawaiian Islands. Many thousand of them over-winter in Hawai'i and return north to Alaska to breed on the arctic tundra. They arrive from Alaska in a drab, frumpy plumage and depart in their breeding finery, a harlequin-plumaged bird with a rust-colored back and bright orange legs. Turnstones have catholic tastes. Flocks of turnstones roam through the bird colonies flipping over rocks,

Ruddy Turnstones bathing in a tide pool at Laysan Island.

SHOREBIRDS OF HAWAI'I

LIMOSA

KIOEA

'ULILI

KŌLEA

'AKEKEKE

HUNA-KAI

Migratory shorebirds of Hawai'i. Pen and ink drawing by the author, 1984.

flies off. The Wandering Tattler is another Alaskan breeder that flies over the North Pacific to winter on the reefs of Hawai'i. The tattler frequents the surf zone of rocky headlands and boulders, gleaning invertebrates from the rocks and sand. Tattlers occur singly, and probably only several hundred winter throughout the Northwestern Hawaiian Islands. In Alaska, the nervous tattler constantly "bobs" as if ready to jump into the air should a predator hit, but in Hawai'i it is relatively relaxed. Not so its cousin, the Sanderling, which always seem to be running. *Huna-kai*, its Hawaiian name, means "sea foam" and no better name could be applied. Flocks of white Sanderlings dash in unison along the edges of retreating waves, probing in the sand for crustaceans. When Jack Kerouac wrote that shorebirds are everlastingly tuned to the pitiless sea, he was no doubt referring to Sanderlings.

Several thousand Pacific Golden-Plovers winter in the Leeward Islands. Plovers arrive in Hawai'i from Alaska after a 3,000-mile flight taking only several days in August. Lacking webbed feet, they are unable to sit on the water and must travel nonstop. The exhausted birds return to specific areas and set up winter territories, which they defend against newcomers. *Kōlea*, as they are called in Hawaiian, scatter throughout the Islands in grass and open habitats to forage on insects and crabs.

seeking insects and carrion. Aside from turning over a few stones, they also attend the birth of monk seals, awaiting the placenta delivery to feast on the blood-rich meal and the attendant flies. Turnstones also peck open scabs on wounds of monk seals, evidently to have a blood meal now and then. They can even break open seabird eggs with their stout, chisel-like bills. Like jackals following the lions, turnstones also scrounge tidbits from the egg meals of curlews.

It is difficult to convey the charisma of a mousy gray bird, but the Wandering Tattler is a charmer. The bell-like tinkling of its call is the basis of its charm for me. *'Ūlili* is its Hawaiian name, an onomatopoeic rendition of the call it gives when it

References

1. Ziegler (1990a).
2. E. H. Bryan Jr. (1954).
3. Clapp and Wirtz (1975).
4. Barratt (1987).
5. Barratt (1987).
6. Buck (1953).
7. Lisiansky (1814:251–256).
8. *The Polynesian* (1844:87–91).
9. Clapp and Wirtz (1975).
10. *The Polynesian* (1857:40).
11. *The Friend* (1872:31).
12. Clapp and Wirtz (1975).
13. Personal records of the Schlemmer family (rent receipts from Public Lands Office, Honolulu), Waimānalo, Hawai'i.
14. Clapp and Wirtz (1975).
15. Personal records of the Schlemmer family (transcription from *The Honolulu Advertiser*, 21 September 1904:2), Waimānalo, Hawai'i.
16. Farrell (1928:399).
17. Farrell (1928:397–398) in Olson and Ziegler (1995).
18. E. H. Bryan Jr. (1942).
19. Ziegler (1990b).
20. Elschner (1925:56).
21. Clapp and Wirtz (1975).
22. Ziegler (1990b).
23. Olson and Ziegler (1995:120).
24. Olson and Ziegler (1995:122–123).
25. Lisiansky (1814) in Apple (1973:64).
26. Ziegler (1990b:44).
27. Marks (1995a).
28. Marks (1995a).
29. Marks and Hall (1992).

17

Pearl and Hermes Reef

Kōlea kau ʻāhua, a uliuli ka umauma hoʻi i Kahiki.

"Plover that perches on the mound, waits till its breast darkens, then departs for Kahiki."
The darkening of the breast is a sign that a plover is fat.
Applied to a person who comes to Hawaiʻi, acquires wealth, and departs.

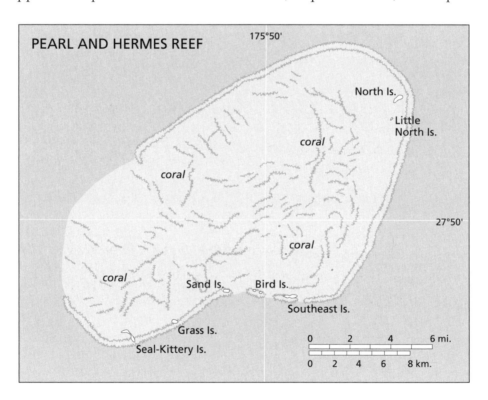

PEARL AND HERMES REEF

175°50'

North Is.

Little
North Is.

coral

coral

27°50'

coral

coral

Sand Is.   Bird Is.

Southeast Is.

Grass Is.

Seal-Kittery Is.

| 0 | 2 | 4 | 6 mi. |

| 0 | 2 | 4 | 6 | 8 km. |

Map of Pearl and Hermes Reef.

Like billboard blimps, the turquoise under-bellies of cumulus clouds advertise the location of Pearl and Hermes Reef, situated about 165 miles northwest of Lisianski and 75 miles southeast of Midway Atoll. Similar to French Frigate Shoals but without a pinnacle, Pearl and Hermes Reef is a true atoll that is primarily underwater, fringed with shoals and sand islets, and open to the ocean on the west. The submerged platform is estimated to be about 20 miles long and 12 miles wide with about 306 square miles within the 100-fathom line.[1] Pearl and Hermes Reef has only seven islets above sea level today. Sand, Bird, Planetree, and Seal-Kittery are bare islets while Grass, Southeast, and North are tufted with coastal shrubs and beach grass. Southeast Island is the largest in the atoll with 34 acres. The emergence and subsidence of islets proves the atoll to be a dynamic system. Over time, the estimates of dry land have varied between 83 and 139 acres.[2]

It is one of those quirks of fate, foreshadowing future events, that when the English whale ship *Pearl* ran aground on uncharted reefs near Midway, she hit the only pearl beds in the entire territory of the United States. In these waters to pursue whales with her sister ship *Hermes*, she hit hard on 25 April 1822. *Hermes* attempted to render aid, but both met with disaster.[3] After the *Pearl* and *Hermes* crews assembled on shore, James Robinson, carpenter's mate, supervised the building of a small vessel named the *Drift* out of the combined wreckage. The castaways evidently made it back without the loss of life, and Robinson remained in Hawai'i. King Kamehameha II helped Robinson establish a chandlery and shipyard at Honolulu Harbor, building ships for Hawaiian royalty. When Robinson died, he left a half-million-dollar estate.[4]

Claimed by King Kamehameha III in 1854, the atoll was not surveyed until 1857 when the *Manuokawai* passed Pearl and Hermes Reef and saw six islets "located some distance inside of the reef, in what seemed to be a large lagoon, and abounded with birds, seal and turtles; no safe anchorage outside the reef . . . 1053 miles from Honolulu."[5] In 1859, Captain N. C. Brooks of the *Gambia* took possession of the atoll on behalf of the United States. Brooks reported that the atoll consisted of twelve islets, though his unpublished chart shows only eleven. The ship *Laconda* made the first thorough survey in 1867 and produced a reliable chart of the wide maze of reefs.[6]

Because of its small land base, this atoll was

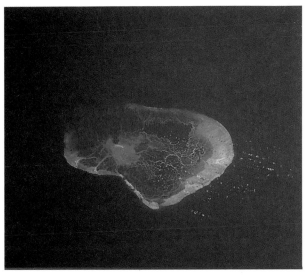

Satellite view of Pearl and Hermes Reef. (NASA photo)

largely spared the ravages of the guano miners and feather hunters. However, in 1904, the Japanese schooner *Wiji Maru* crashed into the reef carrying $20,000 worth of feathers. The ship was wrecked but the crew was saved. The *Thetis* stopped off regularly and never found any Japanese, but in 1916, First Lt. W. H. Munter noted rabbits on Southeast Island as well as the thatched huts of Japanese fishermen.[7] The Tanager Expedition spent a few days here in 1923 and killed all but one rabbit on Southeast Island. The expedition transplanted several kinds of beach vegetation, in addition to discovering two new islets. In 1927, Captain William G. Anderson found pearls in the Black-lip Pearl Oysters here. Anderson founded the Hawaiian Sea Products Company, using the schooner *Lanikai*, formerly a German trading vessel, ironically named *Hermes*.[8]

Within 3 years of the discovery, about 150 tons of oyster shells had been removed and shipped to button manufacturers in San Francisco and New York.[9] The Bureau of Fisheries then thought they had better survey the situation. The U.S. Navy loaned the minesweeper *Whippoorwill*, sister ship of the *Tanager*, for the expedition to Pearl and Hermes Reef. They got under way on 15 July 1930. For the 5-week stay, forty drums of gasoline, a ton of scientific equipment, several tanks to hold live oysters, and 17 tons of ice were brought. Systematic surveys of the reef's position and construction, invertebrates, and fish populations were conducted, but special attention was paid to the location of the pearl oysters. Paul Galtsoff of the Bureau brought along Filipino divers, who were familiar with the atoll, to retrieve oysters for study. The divers collected only 470 oysters, found wedged in the bases of tall coral candelabras. Only 10 percent of the oysters contained pearls. Captain Anderson reportedly had found 20,000 pearls (the largest valued at $5,000!), so Anderson must have taken over 200,000 oysters in the 3 years since discovery of the resource. Dr. Galtsoff concluded that the oyster beds were badly depleted and if not protected from future exploitation the species would cease to exist. During 2 days in 1950, eight divers found six oysters, and in 1969 only one live oyster was seen.[10]

Aerial view of Pearl and Hermes Reef.

On 22 December 1952, Pearl and Hermes Reef again made the news when the freighter SS *Quartette* became the fourth known ship to run afoul of the reef. The vessel, transiting with 10,000 tons of animal fodder and forty-four crewmen from Galveston, was on its way to Korea when it ran afoul of the reef in heavy seas and 35-knot winds. The tug *Ono* came from Honolulu to rescue the 7,000-ton liberty ship but failed to do so. The ship and crew spent Christmas stuck fast to the reef before they were taken to Midway.[11] The ship broke up and has since provided a slope for sea turtles to bask upon.[12] More recently, a racing yacht ran into the reef and its broken pieces are still present.

The proximity of Pearl and Hermes Reef to Midway has invited military trespass over the years. During World War II, the Navy and Marines strafed and bombed the island. An illegal amphibious landing on Southeast Island in the early 1960s brought seeds of Field Mustard imbedded in the mud affixed to the tracked vehicles from Midway. U.S. Fish and Wildlife Service personnel used herbicide and hand pulling to successfully control it.[13] In the 1980s, Navy helicopters on patrol visited the islets. The choppers hovered at 20 feet (they had been warned not to land) while crewmembers tried to procure glass balls with a long-handled dip net.

About 160,000 breeding birds of seventeen species breed on the islets of Pearl and Hermes Reef. Over 20 percent of the world population of Black-footed Albatrosses nest here. They prefer the sandy areas between the shrubs and grasses. Pearl and Hermes Reef is also an important nesting colony for Tristram's Storm-Petrels.[14] These tiny cousins of the albatross dig

T. Wilson holding a Japanese skull discovered on Pearl and Hermes Reef in 1923. (Bishop Museum photo)

burrows under the vegetation that keeps the sand from blowing away. The only recorded Hawaiian nest site for the Little Tern is at Pearl and Hermes Reef.[15] A pair of this Australasian species has nested here several times in recent years and successfully raised young on Southeast Island. Perhaps these are founders of a new colonization, a process that happens less frequently today than in the past as a result of the overall decline in animal populations.

As mentioned in chapter 2, 108 Laysan Finches were released here in 1967. In less than 30 years, the beak of Pearl and Hermes finches shortened, demonstrating that selective changes in a trait can occur in only several generations in response to the local environment. However, the candidates must be genetically flexible; island songbirds are more likely than pantropical seabirds to exhibit genetic changes in several generations. The seabirds' wide geographic range makes genetic isolation problematic.

Laysan Ducks were also taken to Pearl and Hermes Reef and released on 18 March 1967. The first two ducks out of the cage flew out to sea, so the other twelve were wing-clipped before release. The last sighting was of a nesting pair that failed to produce in late September of the same year. Laysan Rails from the Midway population were

15. Conant et al. (1991).
16. Amerson et al. (1974).

Tiger Sharks are common inshore denizens at Pearl and Hermes Reef.

also introduced to Pearl and Hermes Reef in June 1929 by Captain Anderson, but when the atoll was revisited in 1930, only a few seabirds and little vegetation had survived the severe winter storms.[16] Thus, the Laysan Rail was lost at Laysan, Lisianski, Pearl and Hermes Reef, and Midway.

References

1. Amerson et al. (1974).
2. E. H. Bryan Jr. (1942).
3. Amerson et al. (1974).
4. Galtsoff (1931).
5. *The Polynesian* (1857:40).
6. Galtsoff (1931).
7. Amerson et al. (1974).
8. Apple (1973).
9. Apple (1973).
10. Apple (1973).
11. *The Honolulu Advertiser* (1952).
12. Kam (1984).
13. King (1973).
14. Rauzon et al. (1985).

18

Midway Atoll

I wawā no ka noio, he i'a ko lalo.

"When the noddy tern makes a din, there are fish below."
When people gossip, there is reason.

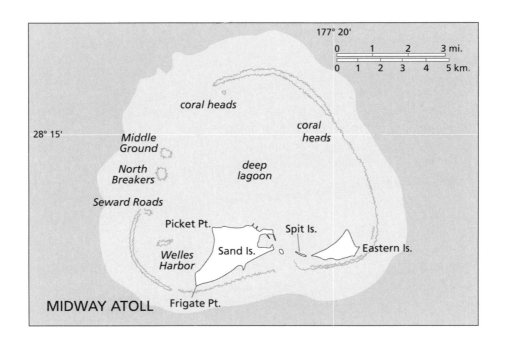

Map of Midway Atoll.

Midway Atoll is the best known North-western Hawaiian Island and often the only one with name recognition. More people have visited Midway than have visited all the other Leeward Islands combined. It is an oasis of green in a blue desert, and the verdure is welcome balm for sun-dazzled eyes. Ironwood trees and Coconut Palms provide cool shade, and a garden of exotic flowering plants, growing on tons and tons of imported topsoil, approximates Eden. Love birds, aka

White Terns, flutter just out of reach and albatrosses waddle with their "Bowery staggers" over every square foot of land. Tropicbirds turn aerial cartwheels over the cerulean lagoon while monk seals roll around on white-sand beaches littered with cowry, mitre, and tun shells.

Midway is roughly equidistant from the Asian and North American continents and halfway around the world from Greenwich, England. The atoll consists of three islands: Sand Island, about 1,200 acres in size; Eastern Island, 334 acres; and tiny Spit Islet that is always changing shape. Considered together, they are the Midway Islands, the legal name on nautical charts. With approximately 1,535 acres, Midway is the largest of the Northwestern Hawaiian Islands. The circular atoll is about 6 miles wide and suggests that Midway was once as large as the island of Lāna'i, perhaps 3,000 feet high and 140 square miles in size.[1] As the original volcanic island, estimated to be approximately 28 million years old, eroded below sea level, the crust of coral built up to 1,261 feet thick over the ancient lava surface. The flanks of the volcanic peak slope down to the abyssal plain of the Pacific Ocean basin, 18,000 feet below sea level.

Midway was discovered on 5 July 1859 by Captain N. C. Brooks of the *Gambia,* outfitted for sealing and searching out guano deposits. He called it Middlebrooks ("Middle" for its location in the Pacific and "Brooks" after himself) and claimed it under the Guano Act of 1856, a law that authorized Americans to temporarily occupy unclaimed Pacific islands to obtain guano.[2] As a result of staking this claim for the U.S. government instead of the Kingdom of Hawaii (a peculiar action in view of the Hawaiian registry of the ship), Midway is the only island in the Hawaiian Archipelago that does not belong to the state of Hawai'i.[3]

Brooks disclosed the location of Middlebrooks to the Pacific Mail Steamship Company, which needed a coal-deposit station on its California-to-China route. The company, in turn, urged

"House and water hole of shipwrecked sailors on 'Upper Brooks,' Midway Islands (uninhabited)," circa 1890. (Hawaii State Archives)

the U.S. Navy to gather more information. The Navy, distracted by the Civil War, did not focus on Midway until 1867. Secretary of the Navy Gideon Welles then directed the steam-powered sloop-of-war *Lackawanna* "without unavoidable delay . . . to take possession of it in the name of the United States, and to make as accurate and complete a survey of it as possible."[4] The wooden-hulled *Lackawanna* was already anchored in Honolulu Harbor, having been deployed to Hawai'i after the Civil War for an "indefinite period," ostensibly to keep an eye on the developing friction between the French and the Kingdom of Hawaii. "The Hawaiian Government, rather than France, was under surveillance," and the intimidating man-o'-war heightened American presence, threatening Hawai'i's sovereignty.[5] *Lackawanna*'s Captain William Reynolds was not shy about his political views toward U.S. annexation. Like Betsy Ross, Reynolds' wife had even sewn a silken American flag to fly over 'Iolani Palace, the royal residence in Honolulu. And because King Lot, the last ruler of the Kamehameha line, was ill and had no heir, a political crisis seemed imminent. The pro-American newspaper *Pacific Commercial Advertiser* put its own spin on the purpose of the ship in 1867: "Her presence here is quite fortunate on this occasion, and forcibly illustrates the propriety of the policy which we have repeatedly urged, that the American Government should always have a war vessel either at or in the neighborhood of Honolulu. Just now, in these 'piping times of peace,' what better occupation for one of 'Uncle Sam's bull-dogs' than to succor his shipwrecked mariners from a barren sand spit where, without this timely assistance, they might have perished."[6]

King Lot told the U.S. minister of finance that the pending reciprocity treaty would not be signed as long as the *Lackawanna* sat in Honolulu Harbor. So the ship was dispatched to search for Brooks' Island. The Hawaiian legislators were relieved when the sloop sailed away and they proceeded to ratify the treaty that eliminated trade barriers between the United States and Hawai'i. Sugar planters then had a duty-free entry into the U.S. market and business boomed in the Islands.[7]

Meanwhile, on 28 August 1867, Captain Reynolds landed with all available men and "took formal possession of Brooks Island, and reefs for the United States." After a twenty-one–gun salute, they hoisted the U.S. flag that Mrs. Reynolds had made, while an agent of the Pacific Mail Steamship Company, who wanted a coaling station built on the island, cheered along. Fireworks in the form of howitzer and small arms target practice followed. Reynolds remarked, "It is exceedingly gratifying to me to have been thus concerned in taking possession of the first island ever added to the dominion of the United States, beyond our shore. . . ." He hoped that "this instance will by no means be the last of our insular annexations. . . ."[8]

Reynolds named the harbor after Secretary Welles and the atoll entrance for Secretary of State William H. Seward, the man who had recently purchased the nearest continental landmass, Alaska,

1,650 nautical miles north, for 6 million dollars. Welles, no doubt pleased with the success of the mission, passed on the documents and charts of the islands to the U.S. Senate Naval Affairs Committee, who recommended that a naval station be established at Midway. The Civil War had demonstrated the political importance of far-flung islands (such as some Caribbean islands that had sheltered Confederates who successfully blockaded Union shipping). On 1 March 1869, the Congress appropriated $50,000 to pay for blasting a channel through the reef at Midway "to afford a safe rendezvous and port of refuge and resort for the naval and merchant vessels of the United States."[9]

The USS *Saginaw* was dispatched from San Francisco on 22 February 1870 to accomplish the task. The side-wheel gunboat, under the command of Lieutenant Commander Montgomery Sicard, arrived at Midway on 24 March and began blasting the reef. By the end of the summer, Sicard wrote, "I cannot help thinking that the amount of time and work required to make this canal were never properly appreciated. . . ."[10] An editorial in *The Friend* (1871) blasted the project: "Through the misrepresentations of interested parties, backed by the recommendations of a naval officer who was either incompetent to judge or who was influenced by unworthy feelings of prejudice and spite against the Government and people of these islands, the North Pacific Mail Steamship Company was led to believe that, by the expenditure of a reasonable sum, a good harbor could be made at Midway Island, a barren sand bank, enclosed by a coral reef. . . . [They] cut a channel 15' x 450' and the $50,000 was expended. At this rate, the work will cost when completed, at least the sum of $1,300,000! It is very much to be doubted whether the U.S. Government will consent to the throwing away of any more money in the useless attempt to make an available harbor at Midway, after the experiences of the past year."[11]

With the appropriation expended, the job

still incomplete, and the weather beginning to turn bad, Sicard steamed for home via a 60-mile detour to neighboring Kure Atoll (then known as Ocean Island) to rescue anyone perchance shipwrecked there. About 3 A.M. on 29 October 1870, the *Saginaw* slammed into the reef at Kure in calm seas. (The saga is recounted in chapter 21 about Kure Atoll.) After the *Saginaw* fiasco, Congress refused to appropriate funds to complete the reef channel, and the Pacific Mail Steamship Company never developed a coaling station at Midway.

Midway was abandoned until November 1886 when the schooner *General Siegel* called. With a full load of oil, fins, and flesh obtained during a 6-month sharking expedition, the crew prepared for the homeward journey. But as they got ready to sail, the albatrosses began laying eggs and the crew postponed their departure so that they could take back a load of eggs, used as a source of photographic albumen. (Eggs were so palatable that the sailors used to drink them raw. The seamen would also rather eat albatross meat than any meat obtainable in Honolulu.)[12]

Killing sharks has always been a risky proposition because it offends Kamohoaliʻi, Pele's older brother, a demigod of great power and the highest of the shark gods.[13] Like the wrathful winds that ravaged Laysan when the Tanager Expedition was present, a November storm packing gale-force winds parted the *General Siegel*'s anchor chains and drove the schooner onto the reef. The eight crewmen abandoned ship and moved into a wooden house built on Eastern Island during the dredging operation. Within a month, a crewman blew off his right hand while fishing with dynamite and died 10 days later. In May 1887, the seven survivors began repairing a lifeboat that drifted ashore from the British ship *Dunnottar Castle,* which had wrecked at Kure Atoll in July 1886. The crew also worked on an old Japanese sampan to sail in tandem. Survivor Edvert Olsen recounted his story in *Mid-Pacific Magazine* in 1918.

As they worked on the sampan, Captain Frank Asberdine, the mate Adolph Jorgensen, and sailor William Brown went to Sand Island in the lifeboat. Only Jorgensen returned. When asked where his shipmates were, Jorgensen replied that they were on Sand Island and that the captain said the crew could depart if they wished. Suspecting foul play, the crew told Jorgensen they would not leave without seeing the captain. When pressed, Jorgensen confessed, "I tell you, boys, the captain killed Brown and I have been keeping him company to prevent him from killing himself. I wouldn't be surprised to find that he has made way with himself." Olsen continued, "We laid alongside of the island and told Jorgensen to go and find the captain, but he went instead to look for birds' eggs. As soon as we saw this we returned to Midway Island [Eastern Island] and got ready to leave. Everything was packed into the Japanese boat, except an axe, some dried fish and a few other articles, which we left for Jorgensen's use."[14]

After they left Jorgensen on Sand Island, he tore up his clothing to make a raft to come to Eastern Island. Jorgensen sneaked ashore, naked but armed with a rifle. While two of the crew were disabling the other boat in the lagoon, Jorgensen confronted Olsen and they tussled until Olsen got the upper hand. "Tie me up and take me with you," Jorgensen pleaded, but the men were afraid to sail away with the murderer. In the predawn hours, the four crew members sailed without Jorgensen for the Marshall Islands with no navigating equipment. After 20 days of subsisting on rainwater and dried fish, they miraculously reached their destination. The local people were very welcoming, which was fortunate because their sampan sank in the lagoon. They lived with the natives for 3 months on a diet of breadfruit, coconuts, and fish.

Some time later, Olsen and another crewmember traveled with locals in a sailing canoe to another island, some 400 miles away. When they got there, the natives proceeded with them to yet another island and found the New Zealand brig *'Ahukai* outward bound from Honolulu, under the command of Captain Macy. He had heard of the missing schooner and took Olsen and the other man to Jaluit in the Marshall Islands in time to meet the *Lillian* bound for Honolulu. Olsen worked as crew for passage alongside survivors of the *Mana,* which had wrecked on another island. After a 41-day voyage, Olsen finally arrived in Honolulu where he reported the murder of his captain. Captain Walker of the *Wandering Minstrel* offered to retrieve the accused Jorgensen but the government declined.[15]

The *Wandering Minstrel* went to Midway anyway the following year and experienced the same vicissitudes of fortune. While anchored in Welles Harbor on 3 February 1888, the 467-ton fishing vessel was hit by a sudden squall and wrecked on the reef. Captain Walker, his wife, and three sons occupied Sand Island, while the crew shared Eastern Island with the castaway Jorgensen. After 3 weeks, six of the men rigged up a boat and set off. They were never heard from again. Later, the ship's mate, John Cameron, recognizing a kindred

Crosses mark the burial sites of victims from the Wandering Minstrel *wreck. Photo by F. G. Walker, 1891? (Bishop Museum photo)*

spirit in the marooned madman, Jorgensen, joined with him and Moses, a Chinese boy, and left in one of the boats in an effort to reach the Marshall Islands. The 2,000-mile trip was successful, but they failed to report the distress of their mates on Midway. Of the remaining men on Midway, one of the first to die was also the greediest, killing up to fifty albatrosses at a time and sharing none. Another man drowned, three died of scurvy, one later died on the way to Honolulu, and one ended up in a lunatic asylum.[16]

Mrs. Walker recounted her harrowing story of survival. She had worn six heavy sea jackets off the sinking ship. Her dog, "little Jessie," was thrown to her as she pulled away in the lifeboat. Little Jessie dug out petrel burrows and they survived by eating the "puppy birds" and albatross eggs whipped in hot water. When hope ebbed, Providence provided a "hen-coop of potatoes" and a cask of fermented rice that floated in as they neared starvation. When one of her boys lay dying of scurvy, his limbs covered with patches of purple, green, and red, a bottle of lime juice was uncovered in the sand and saved his life.[17]

After surviving for 14 months, Captain Walker and his family and crew were finally rescued, quite by luck it turns out. Mrs. Walker's brother had married a Japanese woman who was very fond of her. Mrs. Walker's sister-in-law became ill and as she neared death she consulted a Japanese soothsayer, who told her that Mrs. Walker was still alive and in great distress. The dying woman begged her husband to make every effort to find his sister, and he asked the captain of the schooner *Norma* to keep an eye out for them. On 26 March, the castaways heard the boom of a cannon as the *Norma* announced its arrival, and they were snatched from the jaws of death. The whole party landed safely in Honolulu on 7 April 1889.

Captain Walker again returned to Midway, further tempting fate, when he accompanied the

1891 Rothschild Expedition.[18] The journal entries of the expedition members documented the desolation they experienced at Midway Atoll.

> July 11, [1891] . . . What I am now on is what is known as Sand Island. This island is almost bare and has hardly any birds. . . . Although the island is comparatively large, it is the most desolate place I ever was on. There is hardly any vegetation except a few tufts of grass on the south end, and in rough weather most of the island is under water.[19]

> [July 12, 1891: Sand Island is] almost a mile long, low and sandy, with a few mounds 12 feet high covered with large scrub at one end; at the other end is a patch of grass. . . . It is a very desolate island with a great extent of low-lying sandy ground, which seems to be swept by heavy seas during heavy weather. . . . There is something melancholy about this desolate place . . . the sigh of the wind 'round the house, the wail of the petrels, at any time melancholy, seems even more so . . . a feeling of depression comes over one. . . .[20]

Undaunted by the desolation, isolation, or the ocean depth, the Pacific Commercial Cable Company set out to link Midway with the world. On 26 May 1900, the USS *Iroquois*, a converted tug acquired by the Navy in the 1898 war with Spain, arrived at Midway to take soundings. They discovered Japanese feather hunters on the island. The poaching on Midway prompted complaints from the cable company that spurred President Theodore Roosevelt on 20 January 1903 to put Midway under Navy Department control

via Executive Order 199-A. By April, the first cable workers steamed into Midway lagoon. They surprised the Japanese schooner *Yeiju Maru* and her bird-killing crew, but seemed unable to stop them. On 3 June 1903, the crew of the *Iroquois* found another Japanese schooner at their bloody business and ordered the Japanese to leave within 48 hours. The Japanese balked at the order and questioned the authority of Hugh Rodman, commanding officer of the *Iroquois*. He retorted that the Japanese had broken many laws including "working foreigners on American soil in violation of the alien contract law" and shortened their grace period to 24 hours, threatening to tow their schooner to Honolulu to face charges. The Japanese left under the cover of darkness and Rodman appointed a cable company man as a naval custodian and justice of the peace to prevent wanton destruction of birds and to keep them from being disturbed, except for purposes of food supply.[21]

Cable buildings and compound, circa 1920. (National Archives, Pacific Sierra Region)

Cable deployment went well and by 3 July 1903 a cable connecting Guam, Midway, Waikīkī, San Francisco, and girding the remainder of the globe was in place for the Fourth of July Independence Day message by President Theodore Roosevelt. In 1904, a detachment of marines arrived ostensibly to "protect property and guard the cable employees from marauders who might visit the islands to kill the sea birds"[22] The marines brought the human population to about 100. Midway's population spiked briefly on 16 September 1906, when the Pacific Mail Steamship *Mongolia* ran aground at the western side of Midway. By dumping much of her heavy load and putting the six hundred passengers ashore, she managed to get off the reef with minor damage. The passengers reboarded and she continued on her way. By the end of the decade, the schooner *Julia E. Whalen* and the British bark *Carrollton,* carrying coal from Newcastle, had also wrecked while supplying the cable station.[23]

It was Daniel Morrison, the superintendent of the Pacific Commercial Cable Company, who transformed Sand Island from a barren waste into the garden that it is today. During his tenure from 1906 to 1921, beach grass from San Francisco was planted to help hold the sand. Ironwood trees from Australia were planted to provide windbreaks. Many still stand today, providing habitat for the largest Black Noddy and White Tern colonies in Hawai'i. Morrison imported shiploads of topsoil —approximately 150 metric tons of it, four times a year for a decade! Between 1906 and 1930, 8,100 tons of topsoil from Honolulu were imported to make a 3-acre vegetable garden.[24] Pan American Airways brought an additional 90 tons of soil from Guam. Soil organisms, insects, and weeds were also inadvertently introduced. Sand Island was further domesticated with a milk cow, some pigs, and chickens to supply the cable station with fresh meat. Eastern Island remained more natural even though a small herd of donkeys maintained themselves on the waterless islet. Of the two hundred species of plants on Midway today, over 75 percent are weeds, ornamental shrubs, or exotic vegetables brought to turn Midway into a small piece of O'ahu.

Midway visitors "saw a real paradise, palm trees, ironwood trees, gardens and everything green. That shows that human efforts, properly applied, can convert barren places into places worth living in, and it seems to me the islands scattered between Midway and Hawaii deserve more attention. . . . [I]t would be worth while for any government experiment station or organization to send fauna to make it a place worth living in."[25]

Bishop Museum biologist William Alanson Bryan reported that there were no rats, cats, dogs, or mosquitoes at Midway in 1902, but flies were abundant. He may have been mistaken about cats, however, because in 1906, eleven imported Common Canaries had to be kept in cages until all the cats were killed on the island. In 1910, the yellow songsters were released on Sand Island and persist today, nesting in the Ironwood groves. In 1905, Laysan Finches were also brought to Eastern Island and became established. In 1910, they were transplanted to Sand Island, where they became abundant. Laysan Rails were first released on Eastern Island in 1891 by none other than Captain Walker of the *Wandering Minstrel,* then second officer of the schooner *Kaalokai,* and also became abundant.[26]

Midway was an Eden where almost everything thrived. "It was nearly Eveless Paradise, however, for only the two head men could have their wives on the island. Two women got along nicely, but a third—they found by bitter experience —caused complications. . . . Most of the employees are Scots and it seems that this nationality is the one best adapted to living on lonely localities. I will repeat the statement that they are zealous, industrious, energetic and satisfied with whatever is offered them."[27]

This bucolic scene was not to last. On 10

October 1920, the first airplane flight over Midway occurred. The plane had arrived on a patrol boat, which lifted the floatplane into the lagoon. The plane took off for an aerial reconnaissance, but before it could complete the survey, the plane developed engine trouble and had to return to the ship for repairs. In 1921, Midway served as a refueling station for a group of naval destroyers, and in 1924, a mine sweeper–seaplane tender arrived to complete the aerial surveys. Eight submarines accompanied the ship on maneuvers. In the early 1930s, the Navy held more maneuvers at Midway and landed 750 marines under a smoke-screen cover.[28] In April 1935, a Pan American Airways plane officially landed, carrying enough materials to build and maintain an airport in the middle of nowhere. They also brought a bit of "civilization" to the island. "That Sunday after dark, we gave Midway Island its first movie show. The only man who failed to attend was the cable company's operator on watch. We had brought two movies and sound machines along, one each for Midway and Wake and a number of programs. The feature picture for the evening was 'Red Hot Mamma.' Mr. Lawn, the cable-company engineer, had not seen a movie in six years, and it was rumored that he preferred life at Midway because of the complete absence of feminine society."[29]

It seems the movie broke his will because he shortly later transferred back to civilization. Pan Am used Midway as a refueling stop on the clipper line from San Francisco to Manila, via Honolulu, Midway, Wake, and Guam. Visitors stayed in the one-story "Gooneyville Lodge," entertained by the sound of courting albatrosses pouring in through its many windows.

Pan Am's great four-motor Sikorsky Clipper seaplanes were flying in from Manila and inadvertently transporting all sorts of bugs in the cargo. Midway already had six species each of ants and beetles, five kinds of wasps, ten types of flies, and a host of other insects, including roaches, earwigs, moths, grasshoppers, thrips, lacewings, and springtails.[30] Additional pests that might threaten the sugar industry in Hawai'i had to be stopped, so the Hawaiian Sugar Planters' Association placed the entomologist Fred Hadden on Midway to fumigate air cargo heading east. But insect pests were nothing compared to the invasion brewing in Japan.

Pan American Airways building, circa 1940. (Bishop Museum photo)

References

1. Bailey (1956).
2. Skaggs (1994).
3. Bailey (1956).
4. Cressman et al. (1990:1).
5. Daws (1968).
6. *Pacific Commercial Advertiser* (1867:2).
7. Daws (1968).

Pan American Clipper floatplane at dock. (National Archives, Pacific Sierra Region)

8. Cressman et al. (1990:2).
9. Cressman et al. (1990:2).
10. Cressman et al. (1990:2).
11. *The Friend* (1871:12).
12. Olsen (1918).
13. Cartwright (1923).
14. Olsen (1918:78, 79).
15. Walker and Harrison (1936).
16. Walker and Harrison (1936).
17. Walker and Harrison (1936).
18. Cressman et al. (1990).
19. Palmer in Bailey (1956:14).

20. Munro in Cressman et al. (1990:20).
21. Cressman et al. (1990:5).
22. Bailey (1956:12).
23. *Pacific Commercial Advertiser* (1903).
24. Hadden (1941).
25. Galtsoff (1931:56).
26. Bailey (1956).
27. Bailey and Niedrach (1951) in Bailey (1956:12).
28. Cressman et al. (1990).
29. Grooch (1936:73).
30. E. H. Bryan Jr. (1954).

The Battle of Midway

Pūhā ka honu, ua awakea.

"When the turtle comes up to breathe, it is daylight."
When a person yawns, sleeping time is over; work begins.

Aerial views of Midway Islands, Eastern and Sand, 1942. (National Archives, Pacific Sierra Region)

The U.S. Navy concluded in a 1939 study of national defense that an air base at Midway Island was second in importance only to one at Pearl Harbor. The report recommended that a ship channel sufficient to accommodate a large tender or tanker be dredged and a base be established to maintain two patrol plane squadrons with adequate storage facilities for aviation fuel. The 120-foot channel project to complete the new harbor engaged 225 men.[1] The Hawaiian Dredging Company eventually removed 3 million cubic yards of coral rubble at a cost of 2 million dollars. The dredging deepened Welles Harbor so that submarines could moor. By 1940, construction of the docks was complete, so vessels could land materials directly on shore without using lighter barges.[2] In September 1940, the Midway Detachment of a Fleet Marine Force landed 196 men and "began the arduous task of making camp, and installing the defenses of Midway" —costing 5 million dollars.[3] Midway was made a national defense area by an executive order dated 14 February 1941, and Midway Naval Air Station was commissioned on 1 August 1941. Streets with lighting, telephones, and buildings were soon in place. The officers' quarters, shops, and administration buildings were on Sand Island and the airstrips were built on Eastern Island. The airport had excellent, all-weather hard surfaces and the capacity to fuel three planes simultaneously.[4] From these initial efforts, the defenses of Midway were made strong enough to withstand the assaults of the Imperial Japanese Navy, which occurred a short time later.

Once again, international affairs caught Midway in their wake. Japan, whose nearest possession was only 1,500 miles away from Midway

Aerial view of Sand Island Harbor, 1942. (National Archives, Pacific Sierra Region)

(about as far away as Honolulu), invaded Indochina in 1940, and pushed eastward into the Marshall Islands. (Japan was considered a threat to Midway as early as 1936 when all "orientals" were removed from Midway and replaced with Guamanian workers.) On 7 December 1941, the "date which will live in Infamy," a Japanese raiding party bombed Midway after they had finished at Pearl Harbor. The attack killed four Americans, destroyed the hospital, and hit the seaplane hanger, but the marines shot back and severely damaged the enemy ships.[5] First Lt. George H. Cannon was killed while manning a gun position and was the

first marine in World War II to receive the Congressional Medal of Honor. Midway's school was named in his honor.

The Japanese returned 6 months later, intent on taking Midway, luring the U.S. fleet into battle, annihilating the Navy, invading Hawai'i, and making the whole Pacific a Japanese "lake." Up to this point, the Japanese forces had been very lucky. They had met little resistance to their advances across the Pacific and they appeared to have an attitude of invincibility. But this time they lacked two key ingredients for victory: radar and the element of surprise. At the time of the Midway battle,

Japan had just developed a prototype of radar for their ships, but the United States had rushed radar to Midway a month before the attack.[6]

The United States had also cracked the Japanese code and learned of their attack plans weeks in advance. To test intelligence penetration, the United States faked a message to the effect that the desalination plant on Midway was broken. Two days later, a Japanese message was intercepted, saying that "AF" was low on fresh water. The code was broken, but could the battle commanders trust that it was not a counterespionage trick?

Admiral Chester W. Nimitz, commander in chief, U.S. Pacific Fleet, made the decision to trust that Naval Intelligence was correct in predicting that Midway was the intended target and a diversionary force moving into the Aleutians only a distraction. He recalled a task force from the South Pacific and acting solely on intelligence, sent his entire fleet to lie in ambush northwest of Midway. Fourteen submarines, which were already harassing Japanese supply lines, were also dispatched to the northwest. Rear Admiral Raymond A. Spruance, task force commander of the aircraft carriers *Enterprise* and *Hornet,* was ready to repel them. Rear Admiral Frank Jack Fletcher's carrier, *Yorktown,* quickly repaired after damage sustained in the battle of the Coral Sea in May 1942, was rushed to Midway. Meanwhile, Admiral Isoroku Yamamoto, commander in chief of the Japanese Combined Fleet, arranged his armada of eleven battleships, eight carriers, twenty-three cruisers, and sixty-five destroyers and moved east into the North Pacific to attack Midway Atoll in three groups.[7]

On 3 June 1942, a Japanese convoy was first spotted 450 miles off Midway by pilot Ensign James Lyle. A report soon after by Ensign Jack Reid alerted Midway that the Japanese were on their way and the strength of their forces was formidable. Early the next morning the Japanese launched 100 bombers and 50 fighters from a position 200 miles north of Midway. The U.S. planes on Midway were launched to engage the enemy. At 6:31 A.M., antiaircraft guns opened up on the Japanese air attackers, but Midway was "softened up."[8] Oil tanks on Sand Island and a powerhouse on Eastern were destroyed. Eleven men died and eighteen were wounded, including Hollywood director John Ford, who had arrived a few days before to film the action. He had positioned his camera on the powerhouse and focused his viewfinder on the Sand Island hanger, expecting it to be a logical target. Instead, a bomb blew up the powerhouse and shrapnel caught Ford in the arm. After recovering his senses, he continued filming. (Ford's *Midway,* filmed "live" and in color, with narration by Henry Fonda, won an Oscar for Best Documentary of the war. *Task Force,* filmed in 1949, starring Gary Cooper and Walter Brennan, blended original footage with dramatizations by actors. *Midway,* made in 1976, borrowed heavily from and is less accurate than the other pictures. *Mister Roberts,* with Henry Fonda, James Cagney, and Jack Lemmon, who won an Oscar for Best Supporting Actor for his part, was also filmed in part at Midway.)[9]

The Japanese held the upper hand in their assault until they discovered the U.S. Naval forces positioned to the northwest of them. As the Japanese were arming their planes for a second land strike, they were attacked by Midway pilots. The first pass of bombers entirely missed the Japanese carriers, and the second and third pass also failed to hit any targets. The U.S. pilots dropped over 100 bombs and were almost annihilated in the process. The aircraft carriers *Enterprise* and *Hornet* launched their Dauntless dive bombers, Devastator torpedo planes, and Hellcat fighters, but still suffered great losses from the swarms of Japanese Zeros defending their carriers. The aggressive Zeros, with big red "meatballs" under their wings, tailed the American Hellcats, running them down. Advice from Jimmy Thach, inventor of the Thach Weave—a formation designed to defend against

Men putting out the fire in the bombed seaplane hanger, 1941. (National Archives, Pacific Sierra Region)

tection scored big hits on the Japanese carrier *Kaga,* killing about 800 in a fiery explosion. The dive bombers hit the carrier *Akagi* twice, causing 221 deaths. Three bombs found the *Soryu,* and the burning hulk was finished off by three torpedoes, killing at least 718. The attack occurred when the carrier decks were full of fueled Zeroes loaded with bombs.

Some of the Japanese aircraft followed the American planes back to the carrier *Yorktown.* Several bombs found their mark and brought the *Yorktown* to a dead stop. The "old battle horse" was abandoned as she listed to port, and she was sunk the next day by Japanese torpedoes. By that time, however, the Japanese had lost their fourth carrier, the *Hiryu,* with its 416 men. Kyoichi Furuya of the destroyer *Yugumo* observed: "As the sun went down, the *Hiryu* was burning fiercely, and seemed to scorch the heavens. The sound of explosions in the powder magazines and the fuel tanks were heard. Our aircraft, having no place to land and with their fuel and ammunitions exhausted, went down

attacking Zeros—was to "Fly low against the water, fly as slow as possible, let them shoot at you, and then turn just before the projectiles hit you."[10] In formation, planes literally weaved in and out, in opposite directions, a difficult maneuver to perform if attacked from several angles, but it worked.

The *Yorktown* strike force arrived at the scene in the nick of time. In 4 minutes of bombing, the battle was essentially won. Although it was well defended, the dive bombers with fighter pro-

tragically beside destroyers and cruisers. There was not a single friendly aircraft left. Our fate is a matter of time. We resolved to do our utmost and await the orders of God . ..."[11]

Admiral Yamamoto ordered a surface bombardment of Midway to cover his retreat, and two of his cruisers collided while trying to avoid U.S. submarines.[12] The Japanese did manage to torpedo the *Yorktown,* but paid with the loss of the destroyer *Mikuma* and 650 lives. Both Japanese and Ameri-

J2F2 amphibious bomber destroyed in hanger, 1941. (National Archives, Pacific Sierra Region)

aircraft carriers were lost, two-thirds of their entire fleet, but the United States only lost the flagship carrier *Yorktown.* The Battle of Midway was the turning point in the Pacific War.[15] The Japanese military were greatly weakened and were never able to mount such a naval challenge again. For the remainder of the war they were in a defensive position. The Japanese people were not told of their devastating loss at Midway; in fact, the Imperial communiqué claimed victory! The great sea battle was a boost to American morale still suffering from Pearl Harbor, and in spite of future U.S. setbacks, confidence of a final victory was palpable.[16] The Pacific Theater shifted south and west, and Midway saw no more action, though the base was busy as a stopover point. Many Hollywood stars came to entertain the troops: Bob Hope, of course, and even Frankenstein in the guise of Boris Karloff visited Midway.

cans in lifeboats at sea were still being picked up as late as 19 June by U.S. search planes. The Japanese fared better than the Americans as prisoners of war. Pilot Frank O'Flaherty and radio-gunner Bruno Gaido, having ditched their plane, climbed into their life raft and were captured by crew of the destroyer *Makigumo.* After being interrogated, they had weights attached to their feet and were buried alive at sea.[13]

The infamous Battle of Midway was over in 72 hours. The Japanese had made several tactical mistakes and had suffered a stunning defeat. The human cost of the battle was about 2,500 Japanese and 320 Americans. The Japanese lost many experienced pilots, judged to be the best in the world at fighting at sea and invaluable for training future pilots at home. The United States lost 133 aircraft versus 278 Japanese planes, primarily destroyed on the sunken carriers.[14] Four Japanese

Wheels of war.

By 1950, the wartime garrison of 10,000 men was gone, leaving a caretaker force for aircraft refueling operations. With the advent of globe-spanning bombers, the runways were expanded to 7,900 feet to accommodate the four-engine, radar-armed, pot-bellied Super Constellations. In 1957, 40 million dollars was spent to establish an impenetrable, all-seeing radar screen between Alaska and Midway. The Distant Early Warning System, or DEW line, enforced by airborne early warning squadrons and a squadron of destroyer escort radar pickets, formed the Barrier Pacific.[17] The DEW line lasted until 1965, when more sophisticated equipment was developed. Midway was one of the Navy's vital bases during the Vietnam War, providing services and materiel support to military aviation and fleet units as well as round-the-clock refueling for aircraft and particularly submarines. Up to 3,000 servicemen and their dependents lived on Midway. On 8 June 1969, President Richard Nixon met President Nguyen Van Thieu of South Vietnam on Midway to discuss the Vietnam War troop withdrawal.

It seemed impossible that the day would come when the Midway Naval Air Station would have its own troop withdrawal. But all things must change. Midway was already a de facto wildlife refuge, but military practices put the wildlife in harm's way. In 1988, the U.S. Fish and Wildlife Service (USFWS) developed an agreement with the military to begin joint management of wildlife resources. When the Department of Defense announced plans to decommission Midway in 1992, it was a relatively smooth transition to the Department of the Interior. The Navy committed to clean up Midway before the USFWS took over and spent almost 75 million dollars to do so. Navy contractors removed antenna wires that killed about 5,000 birds during one 6-month period alone.[18] Contractors also removed bright night lights on the aircraft hangers that drew petrels and shearwaters like moths to a flame, tore down derelict buildings, and removed lead paint chips that poisoned albatrosses. Tons of submarine netting were cut up and transported off the island for recycling on the U.S. mainland.

Leaking fuel tanks were scattered all over Midway. The U.S. Environmental Protection Agency required that 40 cubic yards of chemically tainted soil be removed from the island for proper disposal. Another 588,000 cubic yards contaminated with fuels was treated on-island. Over 45,000 gallons of fuel-contaminated ground water was treated by injecting steam into the ground to volatilize fuels and then sucking them out into an oil-water separator. Some contaminated soil is still there, about 5–12 feet below ground surface.[19]

The U.S. Navy transferred "custody and accountability" to the USFWS on 20 May 1996 and finally departed the island on 30 June 1997. The USFWS now manages the Midway Atoll National Wildlife Refuge separately from the Hawaiian Islands National Wildlife Refuge because Teddy Roosevelt excluded Midway from the original Bird Reservation. The USFWS found that partnering with the private sector could open up Midway to the public while allowing the spectacular natural resources to be protected and enhanced. Midway is managed in partnership with the Midway Phoenix Corporation. The corporation has the rights to bring paying customers to experience the magic of Midway in exchange for maintaining the infrastructure. This arrangement fulfills the USFWS goals of providing environmental education and appropriate wildlife-dependent recreation for the public on this remote refuge in an era of decreasing budgets. The Midway Phoenix Corporation has invested over 8 million dollars since 1996 and has pledged to maintain the runway, requiring about half a million dollars per year. This partnership is an example of the privatization of public resources that was ushered in with the Reagan administration.

Midway only changed hands in the federal system; it still did not join the state of Hawai'i. Perhaps that is best, because the state was considering using it as a "prison in paradise." Instead of prison food, there is a French restaurant on Midway now, underscoring the renaissance that has occurred there. Having been off limits to civilians since Pam Am stopped flying there in 1941, historic Midway Atoll is now open for business as an ecotour destination. Birders, fishers, and photographers flock to the islands to see rare birds, hook big fish, and photograph the fantastic spectacle. Military buffs, veterans of the Battle of Midway, and their relatives also visit. The discovery of the sunken USS *Yorktown* rekindled the spirit of the Battle of Midway. Robert Ballard, the renowned undersea explorer who found the wreck of the *Titanic,* discovered the *Yorktown* on a National Geographic Society–sponsored trip in 1998, 56 years after the aircraft carrier sank.[20] Lying on the abyssal plain of the Pacific Ocean, 16,650 feet beneath the surface—a mile deeper than the *Titanic*—the *Yorktown* sat bolt upright with no biological growth on her. The stainless steel work on the vessel was shiny and one could see all the way across the flight deck. Unlike the *Titanic,* most of the *Yorktown* crew survived. The ship then is not a grave marker, but a symbol of one of the grandest moments in American history, when the Navy turned the tide of aggression.

Midway Atoll has absorbed more money and human labor than many lesser-developed nations. The world has changed so much in the last century and Midway has experienced it all. Finally free from the rush of civilization, Midway has retired from active duty. Once a desolate spot in the middle of nowhere, Midway is blossoming with birds.

"You may look at the plover, peewit, and curlew, but must not shoot; . . . It was the closing act of the Roosevelt administration to set apart

Laysan Albatrosses bond for a lifetime.

these eight islands (or groups of reefs and rocks) as a reservation for the protection of native birds. The time may come when swift excursion steam yachts will carry tourists for cruises among the bird islands in America's great ocean park, very much as in New Zealand and Norway the scenic splendors of the sea coast are exploited for the delectation of the man who enjoys Nature in all her pristine grandeur."[21]

References

1. Cressman et al. (1990).
2. Cressman et al. (1990).
3. Bailey (1956:12).
4. Cressman et al. (1990).
5. Cressman et al. (1990).
6. Fuchida and Okumiya (1955).
7. Lord (1967).
8. Cressman et al. (1990).
9. Cressman et al. (1990).
10. Cressman et al. (1990:195).
11. Cressman et al. (1990:140).
12. Warner (1976).
13. Cressman et al. (1990).

14. Greene (1988).
15. Cressman et al. (1990).
16. Warner (1976).
17. Cressman et al. (1990).
18. Fisher and Fisher (1974).
19. Ogden Environmental and Energy Services Co., Inc. (1994).
20. Allen (1999).
21. Hudson (1911:343).

20

"Gooneys in the Scavvies"

Aia a pohā ka leo o ka ʻaʻo, kāpule ke momona o ka ʻuwaʻu i ka puapua.

"When the shearwaters are heard, the petrels are fat even to their tiny tails."
People snared petrels to eat when they heard fledgling shearwaters cry.

Laysan Albatrosses on the Midway golf course, bathing in and drinking from sprinklers.

Midway's human-designed habitats are alive with birds that seemingly nest everywhere. Lawns, even golf greens, once appeared as they do on U.S. military bases the world over, except that on Midway, thousands of albatrosses "played on through." The recent downsizing of the military has enabled Midway to officially join the other Northwestern Hawaiian Islands in the National Wildlife Refuge System, where native species are favored over non-natives. The golf course is now gone and the "gooneys" have finally won the battle for Midway.

Everyone who has been to Midway is impressed with the antics of the "gooney birds," the nickname given to the albatrosses of Midway. Sailors were amused by the apparent clumsiness of the birds when the birds returned to Midway after many months and in the case of juveniles, several years at sea. The juvenile birds, accustomed to landing on liquid, not solids, often crash-landed, belly-flopped, and flipped over until they got accustomed to landing on sand. Most had awkward moments until they got their "land legs," and many sailors could relate to the feeling. However, it was the birds' outrageous breeding behavior that held people's attention. Incessant yammering and wailing struck some sailors as a lovesick symphony, enhanced by the isolation of Midway. In the local folklore, it was okay when the "gooneys" talked to you, but when you understood what they said, it was time to leave.

Midway has the world's largest nesting colony of Laysan albatrosses; just shy of a million birds visit there. The Navy was never at ease sharing "their" island with birds with an 8-foot wingspan that soared over the runways. In April 1941, the large number of albatrosses on Eastern Island so worried facilities managers that the officer-in-charge recommended that the bird hazard be entirely eliminated. The Navy tried to physically remove the birds from the apron of the runways. When entertainer Bing Crosby saw the botched results while on a USO show visit and mentioned it on a radio interview, the Navy immediately hired biologists Karl Kenyon and Dale Rice to address the "albatross problem."[1] They found that 40 percent of all daylight aircraft landings and takeoffs collided with albatrosses during the November peak of egg laying, and the year-round average of bird strikes was 17 percent. In spite of a rather high rate of collisions, damage to aircraft was sustained in only 1 percent of all takeoffs and landings during daylight.[2] No human lives were ever lost nor aircraft caused to crash because of albatross strikes on Midway (but two birds sucked into Navy jets cost 3 million dollars to repair in 1995!).

It was the albatrosses themselves who paid with their lives. An average of twenty-five birds per day were killed in 1957 alone, meaning that five thousand albatrosses per breeding season perished. "It is certainly impossible for over 3,000 men to be policed to the point where unauthorized killing and harassment of the birds is rendered impossible, and these kills not only reduce the population but also frighten the birds and result in desertion of eggs."[3] Birds that deserted their eggs increased the Bird Air Strike Hazard (BASH) potential because without eggs, the "now-unemployed" birds were free to soar around before they abandoned the breeding grounds for the summer. Sand Island was officially a bird sanctuary. A $50 fine accompanied the killing of an albatross. The official posture of the U.S. Navy toward the birds flying over the island was one of benevolence, yet there was constant conflict between albatrosses and the military's interest in maintaining safe air transport. However, the military could be heavy-handed. In 1964, the Navy paved over all the land 750 feet from the centerline of each of the three runways on Sand Island, necessitating the destruction by asphyxiation of 18,000 incubating albatross, 13 percent of Midway's and about 5 percent of the world's population. "Here we have an example of the ethical paradox by which government sanctioned mass killing is permissible, while the same activity conducted on a small scale by individuals is heavily penalized."[4]

This paradox became more deadly in the psychological "pressure cooker" Midway, or any small island, can become: "The entire base was called to general quarters in the middle of one night in 1963 when the Atomic Underseas Weapons compound was discovered to be unsecured and two of

Dachshund and albatross at hydrant on Sand Island, Midway.

Fields of albatrosses on Eastern Island.

cer ordered them dumped at sea. After a bargeload of rotten carcasses floated back to the beach, the crew had to pick them up and bury them. Finally, the command heeded the biologists' suggestions.

Kenyon and Rice closely studied the "dynamic soaring" of albatrosses. The birds sailed in the updrafts, created when winds hit the dunes, and then flew over the runways to gain altitude before circling back and windsurfing the dunes again. The biologists recommended moving the sand dunes away from the runway. By leveling the dunes, the Navy removed the source of the updrafts and the albatrosses could not sail across the

its three seaman guards shot to death inside. Fearing that a Russian or Chinese communist submarine had raided the stockpile, all hands were issued weapons, machine-guns were set up at various locations, including intersections in the residential area, and a search of the island was conducted. Ultimately, the third guard was discovered dead in an old WWII bunker with an M-1, and a thousand rounds of ammunition; he had apparently shot himself with his side-arm after shooting his buddies for killing a gooney bird."[5]

The Navy thought that hiring federal biologists Kenyon and Rice would give it carte blanche to kill the birds. Although the biologists did allow a kill, they did so to prove their contention that lethal control would not work. Over 36,000 birds were killed in 1957–1958 alone, and the program went on for several years.[6] But as the biologists suspected, after an area had been cleared of birds, young albatrosses took over the vacant territories. These colonizing birds were also removed, but still others filled in after them. Dead albatrosses began to pile up around Midway, so the commanding offi-

Laysan Albatross caught in antenna wires at Midway.

airstrips. This action reduced aircraft bird strikes by 80 percent.

Other biologists have studied the albatrosses of Midway, notably Harvey and Mildred Fisher and Hubert and Mable Frings. In 1959, the Frings determined that the albatross harassment campaigns would eventually eliminate the birds from Sand Island. They recommended that the disturbance be kept up for the benefit of the birds, because they hoped the displaced birds would resettle on Eastern Island, Kure, Lisianski, or Laysan (in fact, a bargeload of juveniles was actually hauled over to Lisianski and placed there). The Frings contended that those islands were not at maximum albatross density, and "improving the habitat" by planting grasses and Ironwoods at Laysan and Lisianski and blacktopping parts of Kure would help the albatrosses suffering human persecution on Midway. "It may seem expensive and rather quixotic to try thus remodeling deserted islands, but this could, within possibly only 5–10 years, result in populations of Albatrosses great enough that total elimination from Midway Islands could be accomplished with an increase in the world population of the birds."[7]

Plaster albatross at Midway.

This single-species approach to wildlife management is unthinkable today, when there are multispecies/ecosystem approaches available. But, in the end, the albatrosses won, and a giant plaster albatross stands on the Midway parade grounds pointing to the sky with its beak, a fitting monument to these fabulous birds that prevailed in their war with the Navy.

Periodically from the 1930s and every winter after 1971, Midway was the home of an exceedingly rare seabird. For years, the sailors of Midway watched for the return of "George, the Golden

Gooney," nicknamed for the amber wash on its crown. George was a Short-tailed Albatross—the grandest albatross in the North Pacific—and stood a head taller than the other albatross species. Conspicuous with his massive pink and blue beak, he was banded number 001, allowing people to confirm his identity each year. Laysan and Black-footed Albatrosses tried to interact with George, but he overpowered the smaller albatrosses with his vigorous dancing style. He remained alone even after another Short-tailed Albatross arrived in a different area on Midway. The two presumably never met before George finally disappeared in the late 1980s.

In the 1990s, two more short-tails frequented separate areas again, but they did meet and were observed in courtship dance in 1994. On two occasions, infertile eggs were produced. Unfortunately, the oldest of the two birds disappeared in 1998, a possible victim of the long-line fishery that is known to kill at least two Short-tailed Albatrosses per year in Alaska. About 1,000 of this endangered species breed on Torishima Island, a volcano off Japan's southern coast. They nest on an active volcano and lost part of their colony to an eruption in 1939, so another nesting site would add a measure of protection to this magnificent species, considered a national treasure in Japan. A new colony will eventually become established on Midway as the Japanese population continues to expand. To speed its inception, biologists are trying to attract more short-tails and hatch imported albatross eggs.

The Short-tailed Albatross was previously known as Steller's Albatross, named after Georg Wilhelm Steller, naturalist on Vitus Bering's cruises in the North Pacific. In a combined quirk of nature and fate, most of the species discovered by or named after Steller have become endangered or extinct. Steller's Sea Cow, a giant version of the Manatee, was exterminated. The Pallas Cormorant discovered by Steller vanished. The Short-tailed (Steller's) Albatross and Steller's Sea-Eagle are both

endangered species in Asia. The latest species in trouble are Steller's (Northern) Sea Lion and Steller's Eider, which recently have become threatened with extinction. Only Steller's Jay is abundant, often heard scolding from a North American conifer. Georg Steller and Vitus Bering themselves went "extinct" on Bering Island in the Bering Sea in 1742.

The birds of Midway must contend with a host of invasive species. There about two hundred introduced species of plants that make Midway a green oasis. Unfortunately, many are weeds that harbor pests. In the tall fields of Golden Crown-beard and Field Mustard, mosquitoes and flies reside. These insects are vectors for avian pox virus, which is particularly virulent in Laysan Albatrosses and Red-tailed Tropicbirds. Pox causes lesions on the beaks and faces of young birds and is sometimes fatal, although recovered birds have immunity. Also, the introduced Common Myna from India takes a toll on the tree-nesting terns' eggs.

But it is the Black Rats, introduced during World War II, that have had the worst effects on Midway's flora and fauna. By eating chicks, eggs, and adult birds, rats have virtually extirpated the Bonin Petrels, the most abundant breeding bird in the late 1930s. Rats also forced the Brown Noddies,

Short-tailed Albatross at Midway. (Photo by Maura Naughton)

With Black Rats eradicated from Midway and Kure Atolls, the Bonin Petrel population has begun to recover.

which nest on the ground elsewhere, to nest high in the Ironwoods. Rats had the run of Midway until 1995, when Eastern and Spit Islands were cleared of rats by the U.S. Department of Agriculture's Wildlife Services unit. The biologists systematically laid out grids of snap traps and rat poison across each island, changing the lures regularly to prevent bait avoidance. By the 1996 breeding season, Bonin Petrels were once again landing on Eastern Island, the first time since 1945. The first chick fledged there in 1996 and since then, many more burrows have been observed on Eastern and Spit Islands. The rat eradication program on Sand Island began in 1996 and also appears successful.[8]

Unfortunately, or fortunately, depending on your point of view, the rats ignored the introduced Ironwood trees. The trees grow 70 feet tall, taking over open areas, shedding their needlelike leaves, and blanketing the ground so that other native plants cannot sprout. They also speed shore erosion by replacing beach-binding *naupaka* (once called "scavvies" by Navy sailors), which were already hard hit by Black Rats. Burrowing birds

lost habitat, but tree-nesting birds benefited.

The canopy of the Ironwood forest resounds with the racket of White Terns returning to their eggs. Trusting that gravity won't prevail, White Terns build no nest; instead, they lay a single egg in the nook of an Ironwood, palm tree, signpost, spigot, gutter —anything flat and reasonably secure. The chicks hatch with extra large feet and well-developed claws that allow them to cling tenaciously to their precarious perches. Unfortunately on Midway, many chicks get blown off their perches and are attacked by introduced ants. The voracious insects damage the young, sometimes attaching themselves to the chicks' webbed feet and eyes and even inside their bills.

White Terns appear like immaculate waifs, hovering just out of reach overhead. Few other birds purposely fly so close to humans. Their translucent satin plumage filters the intense sunlight as they flock together overhead, emitting churring growls from their blue black bills. It is easy to see why they are also called Fairy Terns. The dove-sized birds allow close approach on land, following the intruder with their black mascara–rimmed eyes before gingerly lifting off and rustling into flight. Adults appear dainty and frail on land, but viewing them at sea evokes new respect. They artfully cut the air with a swift rowing stroke, flying with direct purpose, so differently than their aerial musings over land.

I have seldom seen them fishing at sea. Presumably, the birds catch fish with a quick snatch and can hold at least twelve fish in their bill at one time. The serrated bill with the tongue pressed against the upper beak holds the fish in place and

White Terns survive on tropical islands, even if predators abound. They are able to evade most rats, cats, and ants by living in trees. (Photo by John Gilardi)

permits the adult to catch several fish on each fishing expedition. The adult gauges the size of fish for the appetite of the chick, producing larger and larger items as the chick grows. Scientists have used this behavior to obtain fish specimens for museum collections. The only known specimen of a 2-inch-long fish was obtained from a White Tern during the 1923 Tanager Expedition. It was named *Gregoryina gygis,* after Herbert E. Gregory, director of the Bishop Museum. An observer wryly noted: "This is an example of the extreme depravity to which scientists will descend to obtain a new species, namely taking food from a little bird."[9] Un-

fortunately for Dr. Gregory, it was later discovered that *Gregoryina* was a late postlarval stage of *Cheilodactylus vittatus,* and the unique nomenclature had to be displaced.[10] (Recently, a new algal species was discovered and named *Dudresnaya babbittiana* after Secretary of the Interior Bruce Babbitt to commemorate his leadership in creating the Midway Atoll National Wildlife Refuge.)

White Terns are mobsters. Once an immature Steller's Sea-Eagle accidentally blew in to Midway from Korea or Japan. It stayed several months in 1979 preying on albatrosses. Although the local terns had never seen this type of bird before, flocks

Extinct Laysan Rail on nest. Rails brought in from Laysan became common on Midway until Black Rats arrived during the war buildup. (Denver Museum of Natural History Photo Archives. All rights reserved)

of White Terns dive-bombed the predator. Other strange birds also elicit this mobbing behavior: egg-stealing Bristle-thighed Curlews are harassed from time to time, a Black Kite from Asia drew a mob of two hundred White Terns in 1998, and a Harlequin Duck, a rare visitor from Alaska, was harassed in the water by terns in 1999. The White Tern, *manu-o-Kū* or bird of the god Kū in Hawaiian, is surprisingly long-lived. A banded individual of known age lived to be 42 years old!

Also scattered in Midway's trees are Black Noddies. Unlike their white relatives, Black Noddies construct a substantial nest from leaves and grasses, cemented with guano to tree limbs. Black Noddies are common on Midway; their presence on other Northwestern Hawaiian Islands is limited by the few available trees. Noddies are nearshore feeders, dipping into the water to catch baitfish and crusta-ceans to feed their single chick. *Noio* is the Hawaiian name for this bird. Because of its propensity to fish near shore and roost at night on land, it was an important guide for the sailing canoes looking for land. As in all matters, there are exceptions, and banding records have shown that Black Noddies fly thousands of miles to remote South Pacific islands. Slightly larger and lighter colored than the Black Noddies are the Brown Noddies, the largest tern nesting in Hawai'i. They usually nest on the ground but have adapted to nesting in trees on Midway to be secure from the ravages of the rats. Such behavioral plasticity has enabled them to be the second most abundant tern in the Hawaiian Islands.

Midway also has the largest nesting colony of Red-tailed Tropicbirds in the archipelago. These ground nesters are hefty birds with an intimidating defense: a sudden loud, grating cry accompanied with a jab of their blood red beak. Like the proverbial dog whose bark is worse than its bite, the tropicbird's stab is rather feeble, especially compared with the silent, but deadly booby jab. The tropicbirds' defense is sufficient, however, to

Red-tailed Tropicbirds fighting for nesting space.

render them less vulnerable than other seabirds to rat predation and they can survive in rat-infested areas. While other birds seek the shelter of a shady bush, flocks of tropicbirds engage in aerial courtship, especially during the heat of the noonday sun. They wheel around each other, squawking repeatedly, switching their two long tail feathers back and forth. Red-tailed Tropicbirds are unique among seabirds because they briefly fly backward, "backpedaling" as it were. Refracted heat waves rising off the hot sands may make their aerial acrobatics possible. Even the cliff-nesting White-tailed Tropicbird occasionally nests in planter boxes or Ironwood trees at Midway.

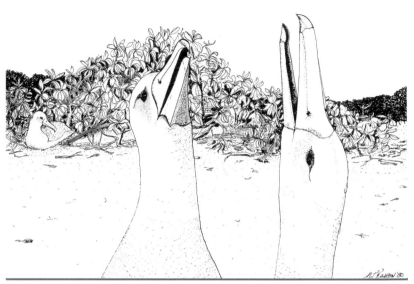

A pair of Laysan Albatrosses performs the "sky-point" behavior of the "gooney" dance.

Many species of waterfowl, shorebirds, and songbirds have stopped by Midway. Individuals of numerous species occur as rare vagrants, particularly during migration. Land bird vagrants, such as the Common Cuckoo that arrived in 1997, are not likely to survive. For most, it is a one-way trip. Seabirds, on the other hand, are more likely to survive. A pair of Little Shearwaters from Australasia once occupied an underground burrow near the Midway golf course. The Jouanin Petrel from the Persian Gulf and the Tahiti Petrel from the Society Islands have also been recorded here.[11]

Nineteen species of cetaceans (whales and dolphins) are widely distributed in tropical waters around Midway. Bottlenose Dolphins and Spinner Dolphins are the most common, inhabiting Midway's lagoon as well as waters of 100 fathoms or less near the other Leeward Islands. A Blainville's Beaked [Densebeak] Whale stranded at Midway in the 1960s. A Cuvier's Beaked Whale [Goose-beaked Whale] stranded there in the 1970s, and another dead Blainville's Beaked Whale washed up

in 1997. I once found an old jawbone from a Sperm Whale. The toothless mandible was all that remained of an individual that washed up on Kure on 8 July 1967. The habitat of these magnificent whales includes the deep waters around these islands, so future strandings of seldom-seen cetaceans are always possible.[12]

Even Elephant Seals are visitors, although rarely. One juvenile female seal, tagged in 1977 at San Miguel Island, California, was seen on Midway in April 1979 lying completely exhausted on the beach. She was covered with Goose-neck Barnacles that had grown on her back during the many months she spent at sea. In her algae-green fur were peculiar bites made by a Cookiecutter Shark, a species of shark that lunges at its intended prey and takes a twisting bite, leaving a round wound as if an ice-cream scoop had been used. This was the first Elephant Seal seen since Captain Morrell reported "sea elephants" from Kure Atoll in 1825. Although reports of "sea elephants and sea leopards" at Pearl and Hermes Reef and Kure were made during the last century, these are now thought to represent misidentified monk seals. Hawaiian Monk Seals have staged a comeback since the military left. In the 1980s, with the military present, only a few seals hauled out. In the late 1990s, as many as eighty have been counted. The public finally has a chance to see the venerable monks in their sanctuary.

References

1. Karl W. Kenyon, 1995, personal communication.
2. Rice (1958).
3. Frings and Frings (1956:14).
4. King (1973:93).
5. King Mallory, June 1999, personal communication (see web page: www.silverlink.net/midway/).
6. Frings and Frings (1956).
7. Frings and Frings (1956:16).
8. U.S. Fish and Wildlife Service biologist Nanette Seto, June 1999, personal communication.
9. Tinker (1978:237).
10. Olson (1996).
11. Seto et al. (1996).
12. Shallenberger (1979).

21

Kure Atoll—Dark Side of the Sun

Mai ka pi'ina a ka lā i Ha'eha'e a i ka lā welo i Kānemiloha'i.

"From where the sun rises at Ha'eha'e (Kumukahi, Hawai'i Island)
to its setting at Kānemiloha'i (Kure Atoll)."

L. L. Kimura, in Juvik and Juvik, *Atlas of Hawai'i,* 3rd ed.

West view from atop Mt. Kure, the 625-foot loran antenna tower on Green Island, Kure Atoll, 1979. Coast Guard buildings and tower have since been demolished and the island is now abandoned.

"Now hear this! All personnel will be restricted to the barracks today." The voice of the commanding officer of the Kure Coast Guard Station boomed over the loudspeaker. He went on to explain that "space debris" was anticipated to enter the earth's atmosphere over Kure. It was 1978 and the sky was falling on the "Dark Side of the Sun" (with apologies to the rock music group Pink Floyd—their album of the same name was hot at that time and that also was the theme on the crew's t-shirts: a statement of isolation at one of the most remote loran stations in the world). It did not seem so safe to be confined indoors when space debris hit because outside you might at least see "incoming"! But orders were orders. To my knowledge, nothing fell to earth that day, but who knows? NASA (National Aeronautics and Space Administration) knows! Somehow, it seemed appropriate that these wide open spaces at the end of Hawai`i are graveyards for wooden ships, steel ships, and rocket ships.

Standing atop Mt. Kure, the 625-foot-tall loran tower, I thought I could make out Midway, 56 miles to the southeast. Below me, the circular atoll lay like a final period

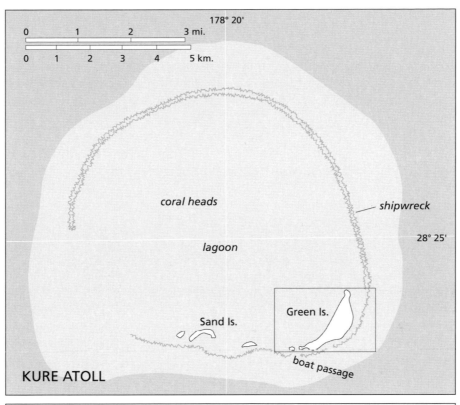

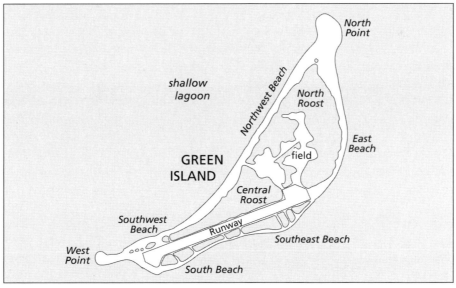

Map of Kure Atoll.

to the sentence that says Hawai'i. The coral reef, about 6 miles across and 15 miles in circumference, is crowned with two islets: Green and Sand. Green Island consists of 236 acres of well-vegetated dunes. It is the only island in the Leewards with a dense forest of Beach *Naupaka.* The crooked stems and rubbery leaves form thickets about 20 feet high that are excellent habitat for boobies and frigatebirds. To the southwest lies 1-acre Sand Island, which is periodically swept by winter storm waves.

No one knows who first wrecked at Kure Atoll. It is possible that the discoverers were Hawaiian voyagers who left Polynesian Rats as a calling card. In 1799, Captain Don M. Zipiani of the Spanish vessel *Senhore del Pilar* discovered an island at about Kure's longitude and latitude and named it Patrocinio. Captain Benjamin Morrell Jr. of the schooner *Tartar* definitely claimed Kure in 1825 and found sea turtles, including two hawksbills, and "sea elephants" in abundance.[1] There are no records available to confirm the alleged discovery of this island by the Russian navigator Captain Kure.[2] We do know that Captain Stanikowitch in 1827 aboard the Russian ship *Moller* may have named the island after Captain Kure, perhaps based on his information of an island reported earlier from this vicinity.[3] The Russian captain/cartographer Krusenstern assumed that Patrocinio, Morrell, Ocean, Massachusetts, Staver, and Cure Islands were all one and the same and synonymized all the names into "Cure Island" in 1835. However, Motu Pāpapapa and Ocean Island became the common names near the end of the nineteenth century, until "Kure Atoll" was formally adopted by the U.S. Board on Geographic Names.

At midnight on 9 July 1837, the British whaler *Gledstanes* was the first of many vessels known to hit the reef at Kure. The *Hawaiian Spectator* reported in July 1838: "Only one man was lost; he jumped overboard, intoxicated. Captain John R. Brown remained on the island till the 15th of December, when himself, with his chief mate

and eight seaman, sailed for the Sandwich Islands, in a schooner they had, with great toil, and perseverance and skill, constructed from fragments of the wreck. The other officers and men who remained several months longer, and endured great sufferings, were subsequently brought off by a vessel dispatched for that purpose by H.B.M.'s Consul at these Islands."[4]

Five years later, in 1842, the New Bedford whaler *Parker* ran aground and four men were lost. The rest clung to the life raft, which took 8 days to reach Green Island. They built a shelter from the wreckage of the *Gledstanes* and survived from day to day. Only a peck of beans and 20 pounds of salt meats had been saved, so the men killed seven thousand seabirds and sixty monk seals for food. A dog from the *Gledstanes* was present and after several weeks it too was captured and eaten. On 16 April 1843, the ship *James Stewart* appeared on the horizon. It landed, picking up the captain of the *Parker* and a chosen few and promised to return as soon as possible. The rest were left provisioned with 20 pounds of bread and 20 pounds of meat per man, plus cloth and other sundries. The castaways did not have to wait long. The passing whaler *Nassau* rescued them within a month.[5] In 1854, a merchant ship from China or the Philippines wrecked on Kure. When Captain Brooks of the Hawaiian bark *Gambia,* after discovering Midway, visited in the summer of 1859, he found the beach littered with pieces of bamboo, China mats, and tubs, but no sign of the castaways. "On the north end I found washed ashore the broadside of a vessel, that had the fore and main channels on from plankshear to below 6 sheets copper. I brought away the copper and door locks, which I found on her cabin doors on the beach. On the stern of a jollyboat I found the name Isaac Holder branded, probably the builder's name. Good water may be obtained on this island. The second island in size is about two miles long and a half mile wide, with little vegetation, few fowls

North Point, Green Island, Kure Atoll.

and plenty of turtles. The third is a mere sand spit."⁶ (The second island mentioned has since disappeared.)

The USS *Lackawanna* surveyed the treacherous reef in 1867 and produced an accurate chart of Ocean Island that should have eliminated shipwrecks at Kure. However, the next wreck was perhaps the most dramatic in the recorded history of Hawai'i. After the frustrating 1870 season of channel dredging at Midway, the USS *Saginaw*, a wooden-hulled sidewheel gunboat, under the command of Lieutenant Commander Montgomery Sicard, headed to Ocean Island to look for marooned sailors. The wind was fair and the engines were running slowly. They expected to near the island at 4 A.M., but at 2:30 A.M. on 29 October, Sicard sensed danger and ordered the engines to be stopped. At 3:30 A.M. the lookout reported breakers ahead and the order to reverse the engines was given. They revved for 10 minutes until a steam connection broke. The *Saginaw* lost control and the currents drove it onto the reef. She careened to port and the smokestack fell over the side. "It was in truth, a remarkable shipwreck. The night had been clear straight, with a moderate breeze. The ship was heading direct for an island whose position and distance—and that a short one—were known, approximately if not precisely. She was making not over two and a half to three knots, yet she ran directly, without any particular

Male frigatebird displaying inflated gular pouch on nau-paka, *Green Island, Kure Atoll.*

that they might have to eat rats if the seals and seabirds departed (which was possibly the same thinking as the original discoverers, who may have introduced Polynesian Rats). The men survived on half rations as they refitted the captain's gig, a light boat with two pairs of oars.

On 18 November 1870, four volunteers—Peter Francis, quartermaster; James Muir; John Andrews; and William Halford, coxswain—under the command of executive officer Lt. J. C. Talbot, set sail for Honolulu. During their 30-day odyssey, they endured three gales that claimed their oars and spoiled their provisions. Exhausted by continual exertion, exposure, and lack of nourishment, they finally reached Kaua'i. What a sight that must have been! But cruel fate was toying with them, because the wind shifted during a heavy rain squall and blew them away from shore. They fought to regain the proper tack into Hanalei Bay only to flounder in the breakers. The gig capsized and Lt. Talbot and two others were washed overboard and drowned. William Halford clung to the boat while James Muir sat in a stupor. Halford "made him fast to the deck" and then carried a tin box containing

lack of vigilance, on a reef which was above water, and on which the breakers were dashing furiously."[7]

At daybreak, the ninety-three members of the crew, plus a Midway dredging party traveling on board, abandoned ship and made it to shore. Little could be salvaged except some lumber to build a house and a boiler that was used to distill fresh water from seawater. Much of the food was spoiled by salt water. To survive, the crew ate albatrosses, but feared that a blow from a wing could break a man's leg. Seals were easily dispatched by a blow to their brittle skulls. "I commenced by sending out parties to kill seal . . . but after about a month I found that, owing to the rapid dimination of the seal, I was obliged to cut the allowance down, and only killed one seal . . . per day for the whole crew."[8] The liver and steaks of the pups were especially tasty. The men of the *Saginaw* thought

Great Frigatebird female tends a newly hatched chick.

A male Brown Booby attends a nest in the interior of Green Island, Kure Atoll.

papers and dispatches to shore. He returned to get the ship's chronometer and returned once again for Muir. At the end of his strength, Halford dragged Muir to shore, after wading through the surf five times. For "showing most heroic fortitude," Halford received the Medal of Honor.[9] Muir died on shore, but Halford got the word out that the rest of the crew was in trouble.

The schooner *Kona Packet* was dispatched, but the schooner's slow speed in winter weather prompted the government to send the steamer *Kilauea* on 26 December to retrieve the marooned crew. On 3 January 1871 at Kure, the *Saginaw*'s carpenter was at work on another boat to carry the crew to Midway when he looked up from his task and saw the smokestack of the *Kilauea*. All hands were rescued and arrived safely in Honolulu on 14 January 1871. In spite of obvious mistakes, Lieutenant Commander Montgomery Sicard was cleared of negligence by a board of inquiry.[10]

On 15 July 1886, the *Dunnotter Castle*, carrying coal from Sydney, Australia, to San Francisco, struck the Kure reef at midnight. The crew were able to save the water casks. Although the contents were tainted with seawater, it was better than the water they found on the island. They fitted the lifeboat, and seven men set sail for the main Hawaiian Islands. During the first 3 weeks at sea, they had but one biscuit and a pint of water per day. For the last 2 weeks, they survived on half that ration! After 52 days, they reached Kaua'i. When word of their arrival reached Honolulu, the steamer *Waialeale* was dispatched to rescue the marooned crew. The ship reached Kure on 20 September, but the island was deserted. After being stranded for 33 days, the crew of twenty-two had been taken off the island by the ship *Birnam Wood* of New Brunswick, Canada, which was heading from Hong Kong to Valparaiso, Chile.[11]

The *Waialeale* did not make the journey needlessly because the ship carried an emissary of King David Kalākaua. King Kalākaua had sent Colonel James Boyd in command of the *Waialeale* to annex the island for the Hawaiian Kingdom. Boyd displayed his commission stating: "I, Colonel James Harbottle Boyd by the power in me vested by His Hawaiian Majesty King Kalakaua's Commission as His Special Commissioner do in His Royal Name take formal possession of Ocean Island or Moku Papapa as a part and portion of His Royal Domain."[12] Boyd saw to it that a house provisioned with essentials, including two 500-gallon water tanks, was erected so that castaways would be able to survive until rescued. Shade tree seeds were also scattered about and three dogs were removed from the island. Within a year, the house blew down and the provisions were rumored to have been pilfered by Japanese feather hunters.[13]

King Kalākaua was inspired by the idea of a confederacy of Polynesian states ruled by the "enlightened, humane and hospitable spirit" of the Hawaiian Government. He helped frame a resolution prohibiting governments from annexing any additional islands in the Pacific. His proposal was largely ignored, and when his emissary approached the U.S. government with the concept that Midway was part of ancestral Hawai'i and should be surrendered, he was summarily dismissed.[14] *The Hawaiian Gazette* (1886) lampooned the annexation of Kure: "As we approached the island, in the ships boat, a venerable turtle who was watching us from a sand bank, arose up on his aft flippers and changing a plug of ship tobacco from his left to his right cheek, and scraping the sand out of his eyes, distinctly remarked *honu*. He then executed a double somersault, into the sea, followed by a double-barreled discharge of plover shot from Dr. Chaddock's fowling piece. . . ."[15]

Kure was acquired as part of the Territory of Hawaii on 7 July 1898 and joined the Hawaiian Islands Bird Reservation in 1909. On 20 February 1936, President Franklin D. Roosevelt placed Kure under Naval jurisdiction by Executive Order 7299. There was no recorded activity at Kure during World War II, but had the Battle of Midway gone as the Japanese hoped, a contingent of 550 men was to have been detailed to occupy Kure Atoll to set up a seaplane and minisubmarine base.[16]

After World War II, President Harry Truman inadvertently returned Kure to the Territory of Hawaii instead of to the U.S. Department of the Interior. Kure Atoll was made a state wildlife refuge under jurisdiction of the Hawai'i Fish and Game Department, part of the City and County of Honolulu, 1,367 miles from O'ahu. In 1961, the State of Hawai'i gave permission to the U.S. Coast Guard to construct a loran station and permanently occupy Green Island. A 4,000-foot runway and a 625-foot-tall loran tower were constructed.

In 1976, while the island was manned by the Coast Guard, the Japanese ship *Houei Maru No. 5* ran aground during a February storm. None of the seventeen fishermen aboard was ever found. The rusty hulk remains on the reef like a tombstone to their memory. Like the *Ocean Pearl,* which wrecked around 1888, and the Japanese steamer *Stato Maru,* which went aground in 1938, many ships have hit this inconspicuous island. There were many causes; bad luck, drunkards at the helm, the island's dubious positions on charts, and sudden violent weather. Kure's treacherous winter weather is due to its location along the southern edge of the Aleutian low-pressure system. The northeast trade winds usually blow 80 percent of the time, especially in the spring and summer. Their strength averages 10 to 15 miles per hour, but in the winter strong west winds often reach gale force.

Temperatures in this subtropical region are relatively mild, ranging from 45 to 90° F, because the surrounding ocean moderates temperature extremes and the near-constant winds have a cooling effect. It can feel like it is going to snow when a damp north wind roars in December. The maritime weather patterns of the North Pacific generally produce rain-bearing storms from December to March. Throughout the Leewards, the low coral islands receive no more rain than the surrounding ocean, usually 30 to 50 inches per year, but the higher basalt islands catch more. Thunder and lightning storms occur periodically over Kure and can create quite a sound and light show. As the sun sets, huge shadows from purple gray cumulus clouds darken the sea. The sky behind becomes salmon and turquoise; the sun burnishes the water. Curtains of rain look as if they are being pulled

In 1976, while the island was manned by the Coast Guard, the Japanese ship Houei Maru No. 5 *hit the reef; no survivors were ever found.*

Changing of the guard at Kure, 1978. The new Coast Guard lieutenant, arriving for his year of command, is met by the outgoing lieutenant on the Dark Side of the Sun.

(3,018 feet down), Suiko (3,540 feet down), and Tenchi Seamount (6,360 feet down) are major seamounts in the Emperor chain that stud the ocean floor (18,000 feet deep) to the western end of the Aleutian Archipelago in Alaska. Along the way, the submarine peaks provide habitat for deep-sea fish. The Pelagic Armorhead, Pelagic Rockfish, and Alfonsino stocks all have drawn foreign fisheries to the area.[18] They, in turn, attract the U.S. Coast Guard, which patrols the region to prevent foreign fishing vessels from entering the 200-mile economic exclusion zone.

from the sea into the sky. A blast of thunder can rumble around the atoll for a full 3 minutes. On moonlit nights, rain squalls create lunar rainbows: double arcs of white light like ghost rainbows on gleaming beaches at midnight where ghost crabs prance.

At 28° 25' N, Kure Atoll is farther from the equator than any other coral reef in the world. Kure's annual coral growth (0.011 inches per year) is just sufficient to offset the rate of annual subsidence (0.0016 inches per year).[17] Beyond Kure, the waters are too cool for coral to grow. One by one, former Northwestern Hawaiian Islands have sunk below the cold North Pacific Ocean. These drowned volcanoes with their ancient coral crowns became the Emperor Seamount chain. Located 162 and 198 miles, respectively, from Kure, Southeast and Northwest Hancock Seamounts are around 780 feet below the surface. The Mellish (60 feet down) and Milwaukee seamounts (1,000 feet down) form the elbow of the submarine mountain chain. Kinmei (180 feet down), Jingu (2,580 feet down), Nintoku

References

1. Woodward (1972).
2. E. H. Bryan Jr. (1954).
3. Apple (1973).
4. *Hawaiian Spectator* (1838:336).
5. Woodward (1972).
6. Brooks (1860) in Woodward (1972:5).
7. *The Friend* (1871:12).
8. Woodward (1972:303–304).
9. Cressman et al. (1990:3).
10. Cressman et al. (1990).
11. Woodward (1972).
12. Department of the Interior letter, 1886–1887, book 28, Hawai'i State Archives, p. 298.
13. Woodward (1972).
14. Daws (1968).
15. *Hawaiian Gazette* (1886:4).
16. Lord (1967).
17. Grigg and Dollar (1980).
18. Gooding (1980).

22

Recognition, Restoration, and Respect

O ka mea i kūpono i kō kākou no'ono'o aku, 'oia kā kākou e mālama.

"What is suitable for us to reflect on is what we should preserve."

A. Fornander, Fornander Collection of Hawaiian Antiquities and Folk-Lore

Hawaiian Monk Seals fighting on Green Island, Kure Atoll.

In 1958, twenty-five monk seal pups were born on Kure Atoll.[1] In 1978, when I arrived, only ten were born. By 1988, the seal population had declined to a few breeding females. Some of the men stationed on Kure used to walk the 2-mile circumference of Green Island daily for exercise and looking for glass balls. The human foot traffic drove many of seals to pup on the more remote Sand Island and the off-limits north point. There, winter waves washed pups into deep water where they became food for Tiger Sharks. The seals were not even safe on Sand Island from glass ball hunters. Navy helicopters hovered above Sand Island while crewmembers tried to scoop up glass balls in long-poled nets, sandblasting monk seals with prop wash. One such helicopter landed on Sand Island, collapsed a dune, rolled over, and wrecked in the surf. Salvage parties removed the valuable parts, but the rotor of the helicopter remains a fit-ting memorial to "glass ball fever."

It is possible that monk seals would have been extirpated from Kure had not William Gilmartin of the National Marine Fisheries Service instituted a "Headstart" program for monk seal pups. In the 1980s, he shuttled under-weight and abandoned female pups from French Frigate Shoals to Honolulu to be fattened up and then to Kure on Coast Guard C-130 planes. His team protected them from disturbance, taught them to fish in an enclosed portion of the reef, and released them when they were deemed healthy. By the time the Coast Guard departed the island in September 1993, there were twenty such imported pups present. Today, the first females released are pupping on their own, helping to return the seal population to former levels before the arrival of the Coast Guard.[2]

The other species of mammal on Kure has

Hawaiian Monk Seal mother and nearly weaned pup lying by the dump on Green Island, Kure Atoll.

ing them with the station .22-caliber rifle were major pastimes at the dump.

The Coast Guard ran the station until advanced communications, available through global positioning satellites, rendered loran obsolete. When the station was closed in 1993, the environmental cleanup began. The tower I once climbed (Mt. Kure) was collapsed and the wires, cables, and lead paint chips were removed from the island and barged back to Honolulu. Many of the cinder block buildings that housed the men were destroyed and the rat-infested dump was bulldozed underground. Exotic trees like Ironwoods and Coconut Palms were cut and fields of the noxious Golden Crown-beard were mowed one last time. The state hired the U.S. Department of Agriculture's Wildlife Services personnel to eradicate the Polynesian Rats and the effort appeared to work. The nights of stomping rats are now only a fond memory and

been extirpated, but its demise was well planned. Kure Atoll was the only Northwestern Hawaiian Island occupied by Polynesian Rats. The presence of rats on Kure suggests that the atoll was visited by Polynesians voyaging from the main Hawaiian Islands or somewhere farther to the south, though it is possible that the rats arrived on the first Western ships to wreck here. Though the rats' diet on Kure was 62 percent vegetable, 30 percent insects, and 8 percent vertebrate flesh, it was the last 8 percent of the diet, of a rat population estimated to have reached 10,000, that was the problem.[3] Rat populations naturally ebb and flow, but the Coast Guard's garbage dump provided a steady food supply and sustained an unnaturally high rat population. Estimated to have a maximum density of seventy-five per acre after their breeding season, rats can climb on and bite adult albatrosses while they incubate their eggs.[4] The genetically naive albatrosses are behaviorally ill-equipped to repel rat attacks. During my stay on Kure, attempts were made to right the wrongs. Stomping rats and plink-

Laysan Albatross fending off a Polynesian Rat that is seeking the bird's egg. Green Island, Kure Atoll, 1978. The rats are now eradicated.

the bird populations are rebounding. Gray-backed Terns and Bonin Petrels have doubled their population size in the absence of rat predation. Like Midway, Kure Atoll is reborn.

On one of my last days at Kure Atoll, I took some Coast Guardsmen out in the Boston Whaler for a final spin. The relaxing color of the aquamarine lagoon was justification enough. But when the speeding boat attracted a pod of Spinner Dolphins to the bow wave, we felt twice blessed. The spinners displayed intense acrobatics by leaping 20 feet out of the water, spinning several rotations, and performing somersaults, back flips, and tail slaps, but we were not content. We stopped the boat and got into the water with them. Instantly, the dolphins changed their behavior and cautiously circled at a distance of about 30 feet. Occasionally, one would shoot past at close range, always coming from behind. The water filled with clicks and whines from the communicating dolphins (no doubt commenting on our clumsiness). Their tolerance and curiosity was touching and their effortless orbiting around us was mesmerizing. Shafts of shimmering gold light added a hypnotic quality, and it appeared to me that our two realities had become one. I thought how these aquatic mammals, with minds not unlike our own, could convey tranquillity. This anthropomorphic projection on my part symbolized to me the connections of all the animals of the Northwestern Hawaiian Islands, where genetically tame wildlife, after two centuries of human insult and injury, still engage us and, in my mind at least, forgive us our trespasses.

After all, a resurrection of righteousness is occurring here. These small islands endured the worst we could dish out: war, murder, exploitation, and pollution. They now are being restored, renewed, and revived. The abandonment of the military bases on French Frigate Shoals, Midway, and Kure; rat eradication on the former bases; weed control; the restoration of Laysan Island; the recovery of seals, sea turtles, and seabirds all represent efforts to rectify wrongs. Even the recent visit to the bottom of the sea to view the USS *Yorktown* in her grave is similar to the journey of the Necker *kūpuna* back to the *ʻāina*. Both voyages recognized and honored antiquity and our ancestors, ancient and modern, who have lived and died in this region. Let us rededicate ourselves, in their memory, with *aloha* for the *ʻāina*.

References

1. Rice (1960).
2. Winning (1998).
3. Woodward (1972).
4. Kepler (1967).

Spinner Dolphins in the lagoon. Note the round Cookiecutter Shark bite in one dolphin on the right. (Photo by John Gilardi)

XXIV-The Island

Goodbye, goodbye, secret island, rose
of purification, navel of gold,
we return, all of us, to the duties
of our mournful professions and
   occupations.

Goodbye, let the great sea protect you
from our barren brutality!
The time has come to hate solitude:
hide, island, the ancient keys
under the skeletons
who'll taunt us til they're dust
in your caves of stone,
our invasion hopeless.

We are home. And farewell,
   squandered and lost,
is just one more, one goodbye
no more solemn than that which lives
   there:
the immobile indifference surrounded
   by the ocean:
a hundred stone faces who gaze
   within
and forever, into the horizon.

     Pablo Neruda,
     from *The Separate Rose*

White-tailed Tropicbird, Phaethon lepturus. *The tropicbird genus is named for Phaethon, a Greek god who one day borrowed the chariot of the sun. When he drove it dangerously close to the earth, Zeus struck him down with a thunderbolt to keep the world from catching fire.*

Appendix: Some Hawaiian, Common, and Scientific Names

Vegetation

Amaranth *(Amaranthus brownii)*

Bamboo *(Schizostachyum glaucifolium)*

Banana *(Musa acuminata)*

Beachgrass *(Ammophila arenaria)*

Beach Morning Glory *(pōhuehue) (Ipomoea pes-caprae)*

Beach *Naupaka,* "scavvies" *(naupaka) (Scaevola sericea)*

Bunchgrass *('emoloa) (Eragrostis variabilis)*

Calcareous Algae *(Halimeda opuntia)*

Coconut Palm *(niu) (Cocos nucifera)*

Common Sandbur *(Cenchrus echinatus)*

Field Mustard *(Brassica campestris)*

Golden Crown-beard *(Verbesina encelioides)*

Goosefoot *(Chenopodium* sp.)

Hairy Horseweed *(Conyza bonariensis)*

Hawaiian Nama *(Nama sandwicensis)*

Indian Pluchea *(Pluchea indica)*

Ironwood *(Casuarina equisetifolia)*

Kukui Tree *(Aleurites moluccana)*

Laysan Sandalwood *(Santalum ellipticum)*

Laysan Sedge *(Mariscus pennatiformis* subsp. *bryanii)*

Morning Glory *(koali) (Ipomoea indica)*

Nihoa Carnation *(Schiedea verticillata)*

Nihoa Palm *(loulu) (Pritchardia remota)*

'Ōhai *(Sesbania tomentosa)*

Pigweed, Purslane *('ihi) (Portulaca lutea)*

Pisonia Tree, Buka *(pāpala kēpau) (Pisonia grandis)*

Puncture Vine *(nohu) (Tribulus cistoides)*

Sea Purslane, Pickleweed *('ākulikuli) (Sesuvium portulacastrum)*

Sweet Potato *('uala) (Ipomoea batatas)*

Tree Heliotrope *(Tournefortia argentea)*

Wiliwili *(Erythrina sandwicensis)*

Invertebrates

Argentine Ant *(Iridomyrmex humilis)*

Black-lip Pearl Oyster *(Pinctada margaritifera)*

Blow Flies (Calliphoridae)

Brine Fly *(Neoscatella sexnotata)*

Brine Shrimp *(Artemia* spp.)

Carrion Beetle *(Dermestes domesticus)*

Centipedes (Geophilidae, Lithobiidae)

Cockroaches (Blattidea)

Conehead Katydid *(Banza nihoa)*

Cowry Shells *(Cypraea* spp.)

Flying Squid (Ommastrephidae)

Ghost Crab *(Ocypode ceratophthalma)*

Giant Devil's Slide Cricket *(Thaumatogryllus* sp.)

Giant Earwig *(Anisolabis* sp.)

Goose-neck Barnacle *(Lepas fascicularis)*

Hawaiian (Black-foot) Limpet *('opihi) (Cellana exarata)*

Hermit Crabs *(Coenobita* spp.)

House Fly *(Musca domestica)*

Land Snails (Endodontidae, Achatinellidae)

Louse Flies (Hippoboscidae)

Octopus *(he'e) (Octopus cyanea)*

Pacific Pelagic Water Strider *(Halobates sericeus)*

Pseudoscorpions (Cheliferidae)

Red Ant *(Pheidole megacephala)*

Rock Crabs *('a'ama)* *(Grapsus* spp.)
Seabird Tick *(Ornithodoros capensis)*
Sea Cucumbers, bêche de mer *(loli)* (Holothuroidea)
Shingle Urchin *(hā'uke'uke)* *(Colobocentrotus
    atratus)*
Shore Cricket *(Caconemobius* sp.)
Silverfish (Lepismatidae)
Spiny Lobster *(ula)* *(Panulirus marginatus)*
Table-top Corals *(Acropora cytherea, A. valida)*
Triton's Trumpet *(pū puhi)* *(Charonia tritonis)*
Weta Cricket *(Deinacrida* sp.)
Wolf Spiders *(Lycosa* spp.)

Fish

Alfonsino *(Beryx splendens)*
Bigeye Tuna *(Thunnus obesus)*
Butterfly Fish *(Chaetodon trifasciatus)*
Conger Eels (Congridae)
Cookiecutter Shark *(Isistius brasiliensis)*
Dolphin Fish *(mahimahi)* *(Coryphaena hippurus)*
Eagle Ray *(Aetobatus narinari)*
Five-horned Cowfish *(Lactoria fornasini)*
Flying Fish *(mālolo)* *(Exocoetus volitans)*
Galápagos Shark *(Carcharhinus galapagensis)*
Gregory's Fish *(Gregoryina gygis,* now *Cheilodactylus
    vittatus)*
Gray Reef Shark *(manō)* *(Carcharhinus amblyrhynchos)*
Jacks *(ulua)* *(Caranx ignobilis, C. melampygus)*
Moray Eels *(puhi)* *(Gymnothorax* spp.)
Pelagic Armorhead *(Pentaceros richardsoni)*
Pelagic Rockfish *(Sebastes matsubari)*
Skipjack Tuna *(aku)* *(Katsuwonus pelamis)*
Squirrelfishes (Holocentridae)
Striped Marlin *(Makaira nigricans)*
Swordfishes (Xiphiidae)
Tiger Shark *(manō)* *(Galeocerdo cuvieri)*
Whitetip Reef Shark *(manō)* *(Triaenodon obesus)*
Yellowfin Tuna *(ahi)* *(Thunnus albacares)*
Zebra Blenny *(pāo'o)* *(Istiblennius zebra)*

Birds

Baillon's Crake *(Porzana pusilla)*
Black Kite *(Milvus migrans)*
Black Noddy *(noio)* *(Anous minutus)*
Black-footed Albatross *(ka'upu)* *(Phoebastria nigripes)*
Blue-gray Noddy *(Procelsterna cerulea saxatilis)*
Bonin Petrel *(Pterodroma hypoleuca)*
Bristle-thighed Curlew *(kioea)* *(Numenius tahitiensis)*
Brown Booby *('ā)* *(Sula leucogaster)*
Brown Noddy *(noio kōhā)* *(Anous stolidus)*
Bulwer's Petrel *('ou)* *(Bulweria bulwerii)*
Christmas Shearwater *(Puffinus nativitatis)*
Common Canary *(Serinus canaria)*
Common Cuckoo *(Cuculus canorus)*
Common Myna *(Acridotheres tristis)*
Egyptian Vulture *(Neophron percnopterus)*
Gray-backed Tern *(pākalakala)* *(Sterna lunata)*
Great Frigatebird *('iwa)* *(Fregata minor)*
Harlequin Duck *(Histrionicus histrionicus)*
Jouanin's Petrel *(Bulweria fallax)*
Laysan Albatross *(mōlī)* *(Phoebastria immutabilis)*
Laysan 'Apapane, Laysan Honeycreeper *(Himatione
    sanguinea freethii)*
Laysan Duck *(Anas laysanensis)*
Laysan Finch *(Telespiza cantans)*
Laysan Millerbird *(Acrocephalus familiaris familiaris)*
Laysan Rail *(Porzana palmeri)*
Little Shearwater *(Puffinus assimilis)*
Little Tern *(Sterna albifrons)*
Mallard *(Anas platyrhynchos)*
Masked Booby *('ā)* *(Sula dactylatra)*
Nihoa Finch *(Telespiza ultima)*
Nihoa Millerbird *(Acrocephalus familiaris kingi)*
Pacific Golden-Plover *(kōlea)* *(Pluvialis dominica)*
Pallas Cormorant *(Phalacrocorax perspicillatus)*
Red-footed Booby *('ā)* *(Sula sula)*
Red Phalarope *(Phalaropus fulicaria)*
Red-tailed Tropicbird *(koa'e 'ula)* *(Phaethon rubricauda)*
Ruddy Turnstone *('akekeke)* *(Arenaria interpres)*
Sanderling *(hunakai)* *(Calidris alba)*

Short-tailed (Steller's) Albatross *(Phoebastria albatrus)*
Sooty Tern *('ewa'ewa) (Sterna fuscata)*
Steller's Eider *(Polysticta stelleri)*
Steller's Jay *(Cyanocitta stelleri)*
Steller's Sea-Eagle *(Haliaeetus pelagicus)*
Tahiti Petrel *(Pterodroma rostrata)*
Tristram's Storm-Petrel *(Oceanodroma tristrami)*
Wake Island Rail *(Rallus wakensis)*
Wandering Tattler *('ūlili) (Heteroscelus incanus)*
Wedge-tailed Shearwater *('ua'u kani) (Puffinus pacificus)*
White-tailed Tropicbird *(koa'e kea) (Phaethon lepturus)*
White Tern, Fairy Tern *(manu o Kū) (Gygis alba)*

Reptiles

Green Sea Turtle *(honu) (Chelonia mydas)*
Hawksbill Sea Turtle *(honu 'ea) (Eretmochelys imbricata)*
Leatherback Sea Turtle *(Dermochelys coriacea)*
Loggerhead Sea Turtle *(Caretta caretta)*
Marine Iguana *(Amblyrhynchus cristatus)*
Pacific Ridley Sea Turtle *(Lepidochelys olivacea)*

Mammals

Black Rat *(Rattus rattus)*
Blainville's Beaked [Densebeak] Whale *(Mesoplodon densirostris)*
Bottlenose Dolphin *(Tursiops truncatus)*
Caribbean Monk Seal *(Monachus tropicalis)*
Cuvier's Beaked [Goose-beaked] Whale *(Ziphius cavirostris)*
Elephant Seal *(Mirounga angustirostris)*
European Hare *(Oryctolagus cuniculus)*
Field Mice *(Peromyscus* spp.*)*
Guinea Pig *(Cavia porcellus)*
Hawaiian Hoary Bat *(Lasiurus cinereus semotus)*
Hawaiian Monk Seal *(Monachus schauinslandi)*
House Mouse *(Mus musculus)*
Humpback Whale *(koholā) (Megaptera novaeangliae)*
Manatee *(Trichechus manatus)*
Mediterranean Monk Seal *(Monachus monachus)*
Norway Rat, Brown Rat *(Rattus norvegicus)*
Polynesian Rat *(Rattus exulans)*
Pygmy Sperm Whale *(Kogia breviceps)*
Sperm Whale *(Physeter macrocephalus)*
Spinner Dolphin *(Stenella longirostris)*
Steller's (Northern) Sea Lion *(Eumetopias jubatus)*
Steller's Sea Cow *(Hydrodamalis gigas)*

Bibliography

Allen, T. B. 1999. Return to the Battle of Midway. *National Geographic* 195 (4): 80–93.

Amerson, A. B., Jr. 1971. The natural history of French Frigate Shoals, Northwestern Hawaiian Islands. *Atoll Research Bulletin* 150:1–383.

Amerson, A. B., Jr., R. B. Clapp, and W. O. Wirtz II. 1974. The natural history of Pearl and Hermes Reef, Northwestern Hawaiian Islands. *Atoll Research Bulletin* 174:1–306.

Apple, R. A. 1973. Prehistoric and historic sites and structures in the Hawaiian Islands National Wildlife Refuge. Report (unpublished), U.S. National Park Service, Honolulu. 111 pp.

Armstrong, R. W. (ed.). 1983. *Atlas of Hawai'i*. University of Hawai'i Press, Honolulu. 238 pp.

Bailey, A. M. 1952. *Laysan and Black-footed Albatross*. Denver Museum of Natural History, Museum Pictorial 6. 80 pp.

———. 1956. *Birds of Midway and Laysan Island*. Denver Museum of Natural History, Museum Pictorial 12. 130 pp.

Bailey, A. M., and R. J. Niedrach. 1951. *Stepping Stones across the Pacific*. Denver Museum of Natural History, Museum Pictorial 3. 63 pp.

Balazs, G. H. 1979. Synopsis of the biological data on the green turtle in the Hawaiian Islands. Final report. National Oceanic and Atmospheric Administration/National Marine Fisheries Service Report No. 79–ABA–02422. Honolulu. 180 pp.

———. 1980. A review of basic biological data on the green turtle in the Northwestern Hawaiian Islands. Pages 42–54 *in* R. Grigg and R. Pfund (eds.), *Proceedings of the Symposium on Status of Resource Investigation in the Northwestern Hawaiian Islands*. Sea Grant Report 80–04. Honolulu.

Barratt, G. 1987. *The Russian Discovery of Hawai'i*. Editions Limited, Honolulu. 256 pp.

Barrett, S. 1996. Disease threatens green sea turtles. *'Elepaio* 56 (6): 41.

Beaglehole, J. C. (ed.). 1967. *The Journals of Captain James Cook on His Voyages of Discovery. The Voyage of the Resolution and Discovery 1776–1780*. Part 1. Cambridge University Press, Cambridge, England. 718 pp.

Beardsley, J. W. 1966. Insects and other terrestrial arthropods from the Leeward Hawaiian Islands. *Proceedings of the Hawaiian Entomological Society* 19: 157–185.

Beckwith, M. 1970. *Hawaiian Mythology*. University of Hawai'i Press, Honolulu.

Berger, A. J. 1972. *Hawaiian Birdlife*. University of Hawai'i Press, Honolulu. 270 pp.

Boffey, P. M. 1969. Biological warfare: Is the Smithsonian really a "cover"? *Science* 163:791–796.

Boggs, C. H., and R. Y. Ito. 1993. Hawaii's pelagic fisheries. *Marine Fisheries Review* 55 (2): 69–82.

Brown, C. A., and V. J. Cabelli. 1964. The susceptibility of birds to tularemia: The Wedge-tailed Shearwater and Black-footed Albatross. Dugway Proving Grounds, Utah. 37 pp. (abstract only).

Bryan, E. H., Jr. 1942. *American Polynesia and the Hawaiian Chain*. Tongg Publishing Co., Honolulu. 253 pp.

———. 1954. *The Hawaiian Chain*. Bishop Museum Press, Honolulu. 71 pp.

Bryan, W. A. 1911. Laysan Island—a visit to Hawaii's bird reservation. *Mid-Pacific Magazine* 2 (4): 303–315.

———. 1931. Life on Crusoe's Isle. *Mid-Pacific Magazine,* pp. 163–165.

Buck, P. H. 1953. *Explorers of the Pacific.* Bishop Museum Special Publication 43.

Butler, G. D., Jr., and R. L. Usinger. 1963. Insects and other arthropods from Kure Island. *Proceedings of the Hawaiian Entomological Society* 18:237–243.

Carlquist, S. 1980. *Hawaii: A Natural History.* SB Printers, Honolulu. 468 pp.

Cartwright, B. 1923. Did Hawaiians know of Necker Island? Science hopes to find an answer. *Honolulu Star-Bulletin,* 21 July, p. 1.

Christophersen, E., and E. L. Caum. 1931. Vascular plants of the Leeward Islands, Hawaii. *Bernice P. Bishop Museum Bulletin* 81. 41 pp.

Clapp, R. B. 1972. The natural history of Gardner Pinnacles, Northwestern Hawaiian Islands. *Atoll Research Bulletin* 163:1–25.

———. 1976. Gray-backed terns eat lizards. *Wilson Bulletin* 88:354.

Clapp, R. B., and E. Kridler. 1977. The natural history of Necker Island, Northwestern Hawaiian Islands. *Atoll Research Bulletin* 206:1–102.

Clapp, R. B., and W. O. Wirtz II. 1975. The natural history of Lisianski Island, Northwestern Hawaiian Islands. *Atoll Research Bulletin* 186:1–196.

Clapp, R. B., and P. W. Woodward. 1968. New records of birds from the Hawaiian Leeward Islands. *Proceedings of the U.S. National Museum* 124:1–39.

Clapp, R. B., E. Kridler, and R. Fleet. 1977. The natural history of Nihoa Island, Northwestern Hawaiian Islands. *Atoll Research Bulletin* 207:1–147.

Clapp, R. B., M. D. F. Udvardy, and A. K. Kepler. 1996. An annotated bibliography of Laysan Island, Northwestern Hawaiian Islands. *Atoll Research Bulletin* 434:1–92. (Available at www.pwrc.nbs.gov/cgi-bin/foliocgi.exe/laysan.nfo/)

Cleghorn, P. L. 1988. The settlement and abandonment of two Hawaiian outposts: Nihoa and Necker Islands. *Bishop Museum Occasional Papers* 28:35–49.

Conant, S. 1985. Recent observations of the plants of Nihoa Island, Northwestern Hawaiian Islands. *Pacific Science* 39:135–149.

———. 1988. Geographic variation in the Laysan Finch. *Evolutionary Ecology* 2:270–282.

Conant, S., C. C. Christensen, P. Conant, W. C. Gagné, and M. L. Goff. 1984. The unique terrestrial biota of the Northwestern Hawaiian Islands. Pages 77–94 *in* R. W. Grigg and K. Y. Tanoue (eds.), *Proceedings of the Second Symposium on Resource Investigations in the Northwestern Hawaiian Islands.* Vol. 1. Sea Grant Report 84-01. Honolulu.

Conant, S., R. B. Clapp, L. Hiruki, and B. Choy. 1991. A new tern *(Sterna)* breeding record for Hawaii. *Pacific Science* 45:348–354.

Conant, S., R. C. Fleischer, M. P. Morin, and C. L. Tarr. 1992. When endangered species are aliens: Some thoughts on the conservation of rare species. *Pacific Science* 46:401–402 (abstract only).

Cressman, R. J., S. Ewing, B. Tillman, M. Horan, C. Reynolds, and S. Cohen. 1990. *"A glorious page in our history." The Battle of Midway.* Pictorial Histories Publishing Co., Missoula, Montana. 226 pp.

Culliney, J. L. 1988. *Islands in a Far Sea: Nature and Man in Hawaii.* Sierra Club Books, San Francisco. 410 pp.

Daws, G. 1968. *Shoal of Time: A History of the Hawaiian Islands.* University of Hawai`i Press, Honolulu. 494 pp.

Dill, H. R., and W. A. Bryan. 1912. Report on an expedition to Laysan Island in 1911. *U.S. Department of Agriculture Biological Survey Bulletin* 42. 30 pp.

Elschner, C. 1925. *The Leeward Islands of the Hawaiian Group.* Reprinted by *Honolulu Advertiser.* 69 pp.

Ely, C. A., and R. B. Clapp. 1973. The natural history of Laysan Island, Northwestern Hawaiian Islands. *Atoll Research Bulletin* 171:1–361.

Emerson, N. 1909. *Unwritten Literature of Hawaii: The Sacred Songs of the Hula. Collected and Translated with Notes and an Account of the Hula.* Charles Tuttle Co., Rutland, Vermont. 288 pp.

Emory, K. P. 1928. Archaeology of Nihoa and Necker Islands. *Bernice P. Bishop Museum Bulletin* 53. 147 pp.

Evering, G. 1980. Who really discovered Nihoa? Report (unpublished), Pacific Science Information Center, Bishop Museum Archives, Honolulu. 4 pp.

Farrell, A. (ed.). 1928. *John Cameron's Odyssey.* The Macmillan Publishing Co., New York. 461 pp.

Fisher, H. I., and P. H. Baldwin. 1946. War and the birds of Midway Atoll. *Condor* 48:3–15.

Fisher, H. I., and M. L. Fisher. 1974. *Wonders of the World of the Albatross.* Dodd, Mead and Co., New York. 80 pp.

Fisher, W. K. 1903. Notes on the birds peculiar to Laysan Island, Hawaiian group. *Auk* 20 (4): 384–397. Quoted in Bailey (1956).

———. 1904. On the habits of the Laysan albatross. *Auk* 21:8–20.

———. 1906. Birds of Laysan Island and the leeward islands, Hawaiian group. *U.S. Fish Commission Bulletin* 23 (3): 767–807. Quoted in Bailey (1956).

Fitzpatrick, S. 1986. *Early Mapping of Hawai'i.* Editions Limited, Honolulu. 160 pp.

Fleischer, R. C., S. Conant, and M. P. Morin. 1991. Genetic variation in native and translocated populations of the Laysan finch *(Telespiza cantans). Heredity* 66:125–130.

Fornander, A. 1916–1920. Fornander collection of Hawaiian antiquities and folk-lore. *Bernice P. Bishop Museum Memoirs* 5 (3).

Freifeld, H. B. 1993. Patterns of vegetation and nesting seabirds at Midway Atoll, Northwestern Hawaiian Islands. M.A. thesis, University of Oregon, Eugene.

Friend, The. 1857. Account of the *Manuokawai.* 6 June, p. 40.

———. 1871. Ocean and Midway Islands. 21 (2): 12–13.

———. 1872. Cruise of the *Kamehameha V* and discovery of the wreck of the North German Brig *Wanderer.* 21 (10): 31.

Frings, H., and M. Frings. 1956. Studies on the albatross of Midway Island. Report (unpublished), Bishop Museum Archives, Honolulu.

Fuchida, M., and M. Okumiya. 1955. *Midway, the Battle That Doomed Japan–The Japanese Navy's Story.* Ballantine Books, New York. 222 pp.

Gagné, W. C., and S. Conant. 1983. Nihoa: Biological gem of the Northwest Hawaiian Islands. *Bishop Museum News* 10 (7): 3–5.

Galtsoff, P. 1931. The U.S.S. Whippoorwill Expedition to Pearl and Hermes Reef. *Mid-Pacific Magazine* 41 (1): 49–56.

Gooding, R. 1980. Exploratory fishing on the Hancock Seamounts by the *Townsend Cromwell: 1976–79.* Pages 241–250 *in* R. Grigg and R. Pfund (eds.), *Proceedings of the Symposium on Status of Resource Investigation in the Northwestern Hawaiian Islands.* Sea Grant Report 80-04. Honolulu.

Greene, J. 1988. *War at Sea.* Gallery Books, New York. 184 pp.

Grigg, R. W., and S. Dollar. 1980. The status of reef studies in the Hawaiian Archipelago. Pages 100–120 *in* R. Grigg and R. Pfund (eds.), *Proceedings of the Symposium on Status of Resource Investigation in the Northwestern Hawaiian Islands.* Sea Grant Report 80-04. Honolulu.

Grooch, W. T. 1936. *Skyway to Asia.* Longmans, Green and Co., New York. 203 pp.

Gup, T. 1985. The Smithsonian's secret contract: The link between birds and biological warfare. *The Washington Post Magazine,* 12 May, p. 8.

Hadden, F. C. 1941. Midway Islands. *The Hawaiian Planter's Record* 45 (3): 179–221.

Harrison, C. S. 1990. *Seabirds of Hawaii: Natural History and Conservation.* Cornell University Press, Ithaca, New York. 249 pp.

Harrison, C. S., T. S. Hida, and M. P. Seki. 1983. Hawaiian seabird feeding ecology. *Wildlife Monographs* 85: 1–71.

Hawaiian Gazette. 1886. Annexation of Ocean Island. 5 October, p. 4.

Hawaiian Spectator. 1838. Ocean Island, map and descriptive note. 1 (3): 336.

He, X., and C. H. Boggs. 1995. Do local catches affect local abundance? Time series analysis on Hawaii's tuna fisheries. Pages 224–240 *in* R. S. Shomura, J. Majkowski, and R. F. Harman (eds.), *Status of Interactions of Pacific Tuna Fisheries in 1995. Proceedings of the Second FAO Expert Consultations on Interactions of Pacific Tuna Fisheries.* FAO Fisheries Technical Paper 365.

Henshaw, H. W. 1918. A mid-Pacific bird reservation. *Mid-Pacific Magazine* 15 (3): 282–285.

Herbst, D. R., and W. L. Wagner. 1992. Alien plants on the Northwestern Hawaiian Islands. Pages 189–224 *in* C. P. Stone, C. W. Smith, and J. T. Tunison (eds.), *Alien Plant Invasions in Native Ecosystems of Hawaii: Management and Research.* CP5U/UH, Department of Botany, University of Hawai'i, Honolulu.

Honda, V. A. 1980. Preliminary results of studies on fecundity of the spiny lobster, *Panulirus marginatus,* in the Northwestern Hawaiian Islands. Pages 143–148 *in* R. Grigg and R. Pfund (eds.), *Proceedings of the Symposium on Status of Resource Investigation in the Northwestern Hawaiian Islands.* Sea Grant Report 80-04. Honolulu.

Honolulu Advertiser, The. 1952. Seas peril 44 aboard MSTS craft. 22 December, p. 1.

———. 1998. Tern Island albatrosses rack up frequent-flier miles. 28 March, p. 1.

Hoogstraal, H. 1972. Birds as tick hosts and as reservoirs and disseminators of tickborne agents. U.S. Naval Medical Research Unit No. 3, 3-TR-62-73 (abstract only).

Howarth, F. G., and S. L. Montgomery. 1998. Insect life in caves. Page 141 *in* S. P. Juvik and J. O. Juvik (eds.), *Atlas of Hawai'i,* 3rd ed. University of Hawai`i Press, Honolulu.

Hudson O. 1911. Uncle Sam's ocean park—a 1500-mile parkway in the Mid-Pacific. *Mid-Pacific Magazine* 1 (3): 339–343.

Humphrey, P. 1965. Smithsonian Year 1965 annual report. Smithsonian Institution, Washington. 30 pp.

Juvik, S. P., and J. O. Juvik (eds.). 1998. *Atlas of Hawai'i,* 3rd ed. University of Hawai'i Press, Honolulu. 333 pp.

Kalmer, A. K., and R. M. Fujita. 1996. Seabird decline in longline fisheries. Report (unpublished), International Union for the Conservation of Nature, by Environmental Defense Fund, Oakland. 8 pp.

Kam, A. K. H. 1984. An unusual example of basking by a green turtle in the Northwestern Hawaiian Islands. *'Elepaio* 45 (1): 3.

Kamakau, S. M. 1961. *Ruling Chiefs of Hawaii.* The Kamehameha Schools Press, Honolulu.

Kenyon, K. W. 1972. Man versus the monk seal. *Journal of Mammalogy* 53:687–696.

———. 1977. Caribbean monk seal extinct. *Journal of Mammalogy* 58:97–98.

———. 1981. *Handbook of Marine Mammals.* Vol. 2. *Seals.* Academic Press, London.

Kenyon, K. W., and D. Rice 1959. Life history of the Hawaiian monk seal. *Pacific Science* 13:215–252.

Kepler, C. B. 1967. Polynesian rat predation on nesting Laysan Albatross and other Pacific seabirds. *Auk* 84:426–430.

Kimura, L. L. 1998. Hawaiian names for the Northwestern Hawaiian Islands. Page 27 *in* S. P. Juvik and J. O. Juvik (eds.), *Atlas of Hawai'i,* 3rd ed. University of Hawai'i Press, Honolulu.

King, W. B. 1973. Conservation status of birds of Central Pacific Islands. *Wilson Bulletin* 85:89–103.

Knudtson, E. P. 1980. Lisianski Island breeding bird population estimates. Report (unpublished), U.S. Fish and Wildlife Service, Hawaiian Islands National Wildlife Refuge. Honolulu. 19 pp.

Krauss, B. 1988. *Keneti: South Seas Adventures of Kenneth Emory.* University of Hawai'i Press, Honolulu. 419 pp.

Kyselka, W. 1987. *An Ocean in Mind.* University of Hawai'i Press, Honolulu. 244 pp.

Ladd, H. S., J. I. Tracey, and M. G. Gross. 1967. Drilling on Midway Island, Hawaii. *Science* 156:1088–1094.

Lipman, V. 1980. Bombs, birds and whales: The little-known story of Kaula. *Honolulu Magazine,* August, p. 50.

————. 1984. The life, death and rebirth of an island. *Honolulu Magazine,* November, p. 86.

Lisiansky, U. 1814. A Voyage Round the World in the Years 1803–1806 . . . in the Ship *Neva.* John Booth, London.

Lobban, C. S., and M. Schefter. 1997. *Tropical Pacific Island Environments.* University of Guam Press, Mangilao. 399 pp.

Lord, W. 1967. *Incredible Victory.* Harper and Row, New York. 331 pp.

Ludwig, J. P., H. J. Auman, C. L. Summer, J. P. Giesy, J. T. Sanderson, J. N. M. DeDoes, and P. Jones. 1996. Reproductive hazards to North Pacific Albatross from PCBs and TCDD-EQS. *Pacific Seabird Group Bulletin* 23 (1): 43 (abstract).

Macdonald, G. A., and A. T. Abbott. 1970. *Volcanoes in the Sea: The Geology of Hawaii.* University of Hawai'i Press, Honolulu. 441 pp.

Marks, J. S. 1995a. Ecology of the Bristle-thighed Curlews in the Northwestern Hawaiian Islands. Ph.D. dissertation, University of Montana, Missoula.

————. 1995b. Laysan Island *Cenchrus* control, 1991–1995: Project review and recommendations for the future. Report (unpublished), U.S. Fish and Wildlife Service, Hawaiian Islands National Wildlife Refuge. Honolulu. 58 pp.

Marks, J. S., and C. S. Hall. 1992. Tool use by Bristle-thighed Curlews feeding on albatross eggs. *Condor* 94:1032–1034.

Marks, J. S., and S. M. Leasure. 1992. Breeding biology of Tristram's Storm-Petrel on Laysan Island. *Wilson Bulletin* 104:719–731.

Miller, W., S. Rosenberger, C. R. Walker, R. L. Corristan, and C. Edwin. 1963. Susceptibility of sooty terns to Venezuelan equine encephalitis (VEE) virus. Army Biological Laboratories, Frederick, Maryland. 17 pp. (abstract only).

Miya, R., and G. H. Balazs. 1993. Ecology and conservation of green turtles in nearshore waters of Waikiki Beach. *'Elepaio* 53 (2): 9–13.

Morin, M. P. 1992. The breeding biology of an endangered Hawaiian honeycreeper, the Laysan finch. *Condor* 94:646–667.

Morin, M. P., and S. Conant. 1990. Nest substrate variation between native and introduced populations of Laysan Finches. *Wilson Bulletin* 102:591–604.

Morin, M. P., S. Conant, and P. Conant. 1997. Laysan and Nihoa Millerbird. Pages 1–20 *in* A. Poole and F. Gill (eds.), *The Birds of North America* No. 302. The Academy of Natural Sciences, Philadelphia, and the American Ornithologists' Union, Washington.

Munro, G. C. 1944. *Birds of Hawaii.* Tuttle and Co., Rutland, Vermont. 189 pp.

Murphy, R. 1967. *A Certain Island.* Avon Books, New York. 239 pp.

Newman, A. L. 1988. Mapping and monitoring vegetation change on Laysan Island. M.A. thesis, University of Hawai'i, Honolulu. 234 pp.

Ogden Environmental and Energy Services Co., Inc. 1994. Base realignment and closure (BRAC) cleanup plan for Naval Air Facility (NAF) Midway Island. Prepared for Pacific Division Naval Facilities Engineering Command, Honolulu. 110 pp.

Olsen, E. 1918. The Midway tragedy. *Mid-Pacific Magazine* 15 (1): 77–81.

Olson, S. L. 1996. History and ornithological journals of the *Tanager* expedition of 1923 to the Northwestern Hawaiian Islands, Johnston and Wake Islands. *Atoll Research Bulletin* 433:1–210.

Olson, S. L., and H. F. James. 1991. Descriptions of 32 new species of birds from the Hawaiian Islands: Part 1, Non-Passeriformes. *Ornithology Monographs* 45:1–88.

Olson, S. L., and A. C. Ziegler. 1995. Remains of land birds from Lisianski Island, with observations on the terrestrial avifauna of the Northwestern Hawaiian Islands. *Pacific Science* 49:111–125.

Pacific Commercial Advertiser. 1859. 18 August. Quoted in R. A. Apple (1973).

————. 1867. *D. Wood* wrecked. 27 April, p. 2.

————. 1885. 28 July, p. 3.

————. 1903. *Julia Whalen* wrecked. 23 October, p. 1.

Parrish, J. L., L. Taylor, M. deCosta, S. Feldkemp, L. Sanderson, and C. Sorden. 1980. Trophic studies of shallow-water fish communities in the Northwestern Hawaiian Islands. Pages 175–190 *in* R. Grigg and R. Pfund (eds.), *Proceedings of the Symposium on Status of Resource Investigation in the Northwestern Hawaiian Islands.* Sea Grant Report 80-04. Honolulu.

Polovina, J. J. 1994. The case of the missing lobsters. *Natural History Magazine* 2:51–58.

Polovina, J. J., and D. T. Tagami. 1980. Preliminary results from ecosystem modeling at French Frigate Shoals. Pages 149–160 *in* R. Grigg and R. Pfund (eds.), *Proceedings of the Symposium on Status of Resource Investigation in the Northwestern Hawaiian Islands.* Sea Grant Report 80-04. Honolulu.

Polovina, J. J., G. T. Mitchum, N. E. Graham, M. P. Craig, E. E. Demartini, and E. N. Flint. 1994. Physical and biological consequences of a climate event in the central North Pacific. *Fisheries Oceanography* 3 (1): 15–21.

Polynesian, The. 1844. *Holder Borden* wrecks. 12 October, pp. 87–91.

———. 1857. Arrival of the *Manuokawai.* Interesting account of her explorations. 6 June, p. 40.

———. 1859. *Gambia*'s account in the leewards. 13 August.

Pukui, M. K. 1997. *'Olelo No'eau: Hawaiian Proverbs and Poetical Sayings.* Bishop Museum Press, Honolulu. 351 pp.

Pukui, M. K., and S. H. Elbert. 1965. *Hawaiian Dictionary.* University of Hawai'i Press, Honolulu. 188 pp.

Pukui, M. K., S. H. Elbert, and E. T. Mookini. 1974. *Place Names of Hawaii,* 2nd ed. University of Hawai'i Press, Honolulu. 289 pp.

Raloff, J. 1996. The pesticide shuffle. *Science News* 149 (11): 1174.

Rauzon, M. J., and K. W. Kenyon. 1982. The inshore diving behavior of the Hawaiian monk seal. *'Elepaio* 42 (12): 107–108.

Rauzon, M. J., C. S. Harrison, and R. B. Clapp. 1984. Breeding biology of the blue-grey noddy in Hawaii. *Journal of Field Ornithology* 55 (3): 309–321.

Rauzon, M. J., C. S. Harrison, and S. Conant. 1985. Status of the sooty-storm petrel in Hawaii. *Wilson Bulletin* 97:390–392.

Regis, E. 1999. *The Biology of Doom: The History of America's Secret Germ Warfare Project.* Henry Holt and Co., New York. 259 pp.

Repenning, C. A., and C. E. Ray. 1977. The origin of the Hawaiian monk seal. *Proceedings of the Biological Society of Washington* 89:667–688.

Rice, D. W. 1958. Birds and aircraft on Midway Island, 1957–58 investigations. U.S. Department of the Interior, Fish and Wildlife Service Special Scientific Report-Wildlife 44:49.

———. 1960. Population dynamic of the Hawaiian monk seal. *Journal of Mammalogy* 4:376–385.

Rice, D. W., and K. W. Kenyon. 1961. The breeding biology of the Laysan and black-footed albatross. *Auk* 38:89–110.

———. 1962a. Breeding distribution, history and populations of North Pacific albatrosses. *Auk* 79: 365–386.

———. 1962b. Breeding cycles and behavior of Laysan and black-footed albatross. *Auk* 79:517–567.

Riley, T. J. 1982. Nihoa Island—an archaeological mystery in the Hawaiian chain. *Field Museum of Natural History Bulletin* 53:21–27.

Rillero, A. 1999. Green sea turtles. *National Wildlife* 37 (4): 41–47.

Rose, R. G., S. Conant, and E. P. Kjellgren. 1993. Hawaiian standing *Kāhili* in the Bishop Museum: An ethnological and biological analysis. *Journal of the Polynesian Society* 102 (3): 273–304.

Rothschild, W. 1893. *The Avifauna of Laysan and the Neighbouring Islands.* R. H. Porter, London. 320 pp.

Scott, G. A. J., and G. M. Rotondo. 1983. A model for the development of types of atolls and volcanic islands on the Pacific lithospheric plate. *Atoll Research Bulletin* 260:11–33.

Seki, M. P., and C. S. Harrison. 1989. Feeding ecology of two subtropical seabird species at French Frigate Shoals, Hawaii. *Bulletin of Marine Science* 45 (1): 52–67.

Seto, N. W. H., J. Warham, N. L. Lisowski, and L. Tanino. 1996. Jouanin's petrel *(Bulweria fallax)* observed on Sand Island, Midway Atoll. *Colonial Waterbirds* 19 (1): 132–134.

Shallenberger, E. 1979. The status of Hawaiian cetaceans. Report (unpublished), U.S. Marine Mammal Commission, MANTA Commission. 103 pp.

Skaggs, J. M. 1994. *The Great Guano Rush: Entrepreneurs and American Overseas Expansion.* St. Martin's Press, New York. 334 pp.

Stephan, J. J. 1984. *Hawai'i under the Rising Sun.* University of Hawai'i Press, Honolulu. 228 pp.

Strazanac, J. S. 1992. Report of June 1990 Nihoa fieldwork, including new arthropod records and W. C. Gagné collections. Report (unpublished), U.S. Fish and Wildlife Service, Honolulu. 12 pp.

Tabrah, R. M. 1987. *Ni'ihau: The Last Hawaiian Island.* Press Pacific, Honolulu. 243 pp.

Tava, R., and M. K. Keale, Sr. 1989. *Ni'ihau: The Traditions of an Hawaiian Island.* Mutual Publishing Co., Honolulu. 137 pp.

Taylor, L. R., and G. Naftel. 1977. How to avoid shark attack (if you happen to be a Hawaiian monk seal). *Oceans* 10 (6): 21–23.

———. 1978. Preliminary investigations of shark predation on the Hawaiian monk seal at Pearl and Hermes Reef and French Frigate Shoals. Final report. U.S. Marine Mammal Commission. NTIS PB 285 626. 34 pp.

TenBruggencate, J. 1994. Rare plants once isles' commonest. *The Honolulu Advertiser,* 13 March, p. 6.

———. 1997. Buoy on guard again to the north. *The Honolulu Advertiser,* 18 November, pp. 1, 7.

———. 1998. Discarded gear damages reefs. *The Honolulu Advertiser,* 10 November, p. 1.

Tinker, S. W. 1978. *Fishes of Hawaii.* Hawaiian Service, Inc., Honolulu. 532 pp.

Udvardy, M. D. G. 1996. Translation of H. H. Schauinsland, 1899, Three months on a coral island (Laysan). *Atoll Research Bulletin* 432:1–53.

United Nations. Group of Consultant Experts on Chemical and Bacteriological (Biological) Weapons. 1970. *Chemical and Bacteriological (Biological) Weapons and the Effects of Their Possible Use. With a special forward by George Wald.* Ballantine Books, New York.

U.S. Congress. House of Representatives. 1969. Committee on Foreign Affairs. Chemical-biological warfare: U.S. policies and international effects. 91st Congress, first session. 2 December. Appendix D. 407 pp.

U.S. Department of the Interior. 1984. Master Plan/EIS, w/technical appendices for Hawaiian Islands National Wildlife Refuge.

Vancouver, G. 1798. *A Voyage of Discovery to the North Pacific Ocean and Round the World, 1790–1795, in the "Discovery" and "Chatham."* Vol. 3. London.

Walker, F. D. 1909. *Log of the Kaalokai.* The Hawaiian Gazette Co. Ltd., Honolulu. 64 pp.

Walker, F. D., and C. A. Harrison. 1936. Wrecked on Midway Islands in 1888. *Paradise of the Pacific,* November, pp. 27–29.

Ware, D. M. 1995. A century and a half of change in the climate of the NE Pacific. *Fisheries Oceanography* 4 (4): 267–277.

Warner, O. 1976. The battle that saved America. *Modern Maturity,* October–November, pp. 11–13.

Wetmore, A. 1925. Bird life among lava rock and coral sand. *The National Geographic Magazine* 48 (1): 77–108.

Wilson, T. 1923. Nihoa—an island long to be remembered by leader of scientists on Tanager. *Honolulu Star-Bulletin,* 13 July, p. 1.

Winning, B. 1998. The roller coaster ride of the Hawaiian monk seal. *Pacific Discovery,* winter, pp. 30–35.

Withers, N. W., R. H. York, and A. H. Banner. 1980. Preliminary notes on growth and toxicity of the dinoflagellate *Gamierdiscus toxicus* from Hawaiian waters. Pages 90–99 *in* R. Grigg and R. Pfund (eds.), *Proceedings of the Symposium on Status of Resource Investigation in the Northwestern Hawaiian Islands.* Sea Grant Report 80-04. Honolulu.

Woodward, P. W. 1972. The natural history of Kure Atoll, Northwestern Hawaiian Islands. *Atoll Research Bulletin* 164:1–318.

Ziegler, A. C. 1990a. Biological observations on Lisianski Island, Hawaiian Islands National Wildlife Refuge, Summer 1990. Report (unpublished), U.S. Fish and Wildlife Service, Honolulu. 30 pp.

———. 1990b. Search for evidence of early Hawaiian presence on Lisianski Island, Hawaiian Islands National Wildlife Refuge, Summer 1990. Report (unpublished), State of Hawai'i, Office of Hawaiian Affairs, Honolulu. 55 pp.

Index

About the Author

Mark J. Rauzon is a biologist, photographer, and author of over twenty science and nature books for children. He has an M.A. in biogeography from the University of Hawai'i and specializes in invasive species management of seabird habitat and island restoration. Rauzon has traveled extensively throughout the tropical and North Pacific Ocean including Alaska.